RELIGIOUS AGRARIANISM AND THE RETURN OF PLACE

SUNY series on Religion and the Environment

Harold Coward, editor

RELIGIOUS AGRARIANISM AND THE RETURN OF PLACE

FROM VALUES TO PRACTICE IN SUSTAINABLE AGRICULTURE

Todd LeVasseur

SUNY
PRESS

Cover images: Permaculture-designed field at Koinonia (top). Permaculture-designed garden plot at Koinonia (bottom). Author's photos.

Published by State University of New York Press, Albany

Printed in the United States of America

For information, contact State University of New York Press, Albany, NY
www.sunypress.edu

Production, Jenn Bennett
Marketing, Anne M. Valentine

Library of Congress Cataloging-in-Publication Data

Names: LeVasseur, Todd, author.
Title: Religious agrarianism and the return of place : from values to practice in sustainable agriculture / Todd LeVasseur.
Description: Albany, NY : State University of New York, 2017. | Series: Suny series on religion and the environment | Includes bibliographical references and index.
Identifiers: LCCN 2017000316 (print) | LCCN 2017044652 (ebook) | ISBN 9781438467740 (e-book) | ISBN 9781438467733 (hardcover : alk. paper) | ISBN 9781438467726 (pbk : alk. paper)
Subjects: LCSH: Agriculture—Religious aspects—Christianity. | Agriculture—Religious aspects—Judaism. | Agriculture—Religious aspects. | Sustainable agriculture.
Classification: LCC BR115.A35 (ebook) | LCC BR115.A35 L48 2017 (print) | DDC 261.5/6—dc23
LC record available at https://lccn.loc.gov/2017000316

10 9 8 7 6 5 4 3 2 1

THIS BOOK IS DEDICATED TO ALL TEACHERS AND ACTIVISTS WHO ARE WORKING ON RETURNING THE HUMAN ANIMAL TO ITS RIGHTFUL PLACE ON THIS SACRED PLANET: GRACIOUSLY LIVING WITHIN LIMITS, WHILE REGENERATING AND REPAIRING WHAT HAS BEEN DAMAGED.

CONTENTS

ACKNOWLEDGMENTS IX

PREFACE XIII

CHAPTER ONE 1
Sustainable Religion, Sustainable Ethics?

CHAPTER TWO 45
Koinonia and Christian Religious Agrarianism

CHAPTER THREE 77
Hazon and Jewish Religious Agrarianism

CHAPTER FOUR 101
The Local ([Farm] Land)

CHAPTER FIVE 137
Concepts of Health

CHAPTER SIX 163
Justice for All: From Soil to Worker, from Individual to Community

CHAPTER SEVEN 181
Conclusion: A Harvest of Ideas

APPENDIX 199

NOTES 203

REFERENCES 227

INDEX 251

ACKNOWLEDGMENTS

Because this is my first solo-authored book, I invite the reader to indulge me over the next few pages as I have many thanks to give.

To begin, I to thank the editorial staff at SUNY Press. This includes Christopher Ahn, Chelsea Miller, Amanda Lanne-Camilli, Jenn Bennett, Laura Poole, Nancy Ellegate, and series editor Harold Coward. I also thank the three anonymous reviewers for their honest feedback that led to a much better book. Any remaining errors are my own.

Because this book is in part a marriage of my personal history bringing together passion and profession, I begin by thanking those on the professional, academic side. First I thank my colleagues in the Religious Studies Department at the College of Charleston (CofC), some of whom I was privileged to learn from as an undergrad and now enjoy as esteemed colleagues. I especially thank Zeff Bjerken for his help as chair and his support of my professional development, followed by thanking our current chair, Elijah Siegler; and then June McDaniel (and Jim) for many edifying discussions over the years. CofC provided me with the opportunity to teach in and then become director of the Environmental and Sustainability Studies Program. I thank Dorian McMillan, Walter Blair, and Ian Rumsey for co-teaching ENVT 200 with me, and Dean Jerold Hale of the School of Humanities and Social Sciences and Dean Mike Auerbach of the School of Sciences and Mathematics for their support of my directorship position. My biggest ENVT thanks goes to Seth Pritchard for his humor, inspiring teaching, and letting me rock out to Michael Franti as we traveled around Estonia for two weeks. Most recently at CofC I have had the honor to become director of our Quality Enhancement Plan (QEP) on Sustainability Literacy. I thank Provost Brian McGee for this opportunity and Divya Bhati and others in the Office for Institutional Effectiveness and Strategic Planning (and all the QEP subcommittee chairs and members) for their collegiality in helping me ably perform these administrative duties. Other colleagues at CofC to thank for being inspirations and friends include John Rashford, Lee Irwin,

Tracy Burkett, Jen Wright, Martin Jones, Simon Lewis, Kristi Brian, Anthony Greene, John Huddleston, Louise Doire, Meg Cormack, Matthew Cressler, Tim Callahan, Ned Hettinger, Reid Wiseman, and Ade Ofunniyin, among many.

My academic career is built on undergraduate classes at CofC and has significantly benefited from a variety of subsequent learning environments at other institutions. In 2001 I was able to participate in a semester of postgraduate study at Murdoch University in Australia. I thank the private benefactor of the Snailwood Trust; the students and staff of the Institute for Sustainability and Technology Policy; the De Silvas for their hospitality and for introducing me to Harry Potter; Patsy Hallen for inviting me and sharing her ecofeminist wisdom; my classmates in Patsy's class; Stephan Harding for coteaching; and a big thanks to Paul Pule for his continued inspiration.

In 2004–2005 I was part of a fantastic learning cohort at the Centre for Human Ecology in Edinburgh, Scotland. I thank all my professors there who helped guide me through my M.Sc., and a very big thanks to Chris Adams for sharing a flat off the Meadows and taking lead on formulating needed adventures around Edinburgh and Mike Price for many a fun table tennis battle. Last, a huge thanks to Alastair McIntosh for the positive impact he has made on my own path.

My entry into the scholarly discussions about religion and the natural world have benefited from the Religion and Ecology, Religion and Food, and Religion and Animals groups of the American Academy of Religion. It has also benefited from pertinent and engaging conferences held by the International Society for the Study of Religion, Nature and Culture. More important, from 2006 to 2010 I matriculated into the Religion and Nature Ph.D. track in the Religion Department at the University of Florida (UF). The learning opportunities presented by this department and the university overall were foundational in shaping this book. For their time, wisdom, guidance, and knowledge, I thank David Hackett and Manuel Vasquez. Helpful leads on Judaism and the environment were offered by Gwynn Kessler and Leah Hochman. For their service and helpful feedback on the project that was the origin of this book, big thanks go to Mickie Swisher and Whitney Sanford. Two other faculty mentors at UF stand out in terms of shaping my scholarly approach to religion and nature. The first is Bron Taylor, whose classes challenged my learning and whose support of my career has been instrumental to my success. I hope we can surf together one day soon! The other is Anna Peterson, who deftly provided needed feedback, important leads, hard-earned wisdom, and generous encouragement at many stages during the writing of this book. Thanks so much, Anna! Last, I formed a number of friendships with members of my cohort—too many to name, so here

I offer thanks to Jimi and Lynne for hearty meals and graceful wisdom; Mike Lemons for fun surf trips that got me out of Gainesville; and Joe Witt, Lucas Johnston, and Bernard Zaleha for continued insights into religion and nature.

Since earning my Ph.D. in 2010 I have been lucky to participate in two summer research adventures that affected how I think about religion and thus this book. The first was a summer fellowship with the Pluralism Project at Harvard University in 2010. I thank Diana Eck, the great staff affiliated with the project, and my cohorts for a memorable time that embodied the academic pursuit of understanding pluralism and dialogue. The second was in 2014 when I was selected to participate in an NEH-funded Summer Institute devoted to Problems in the Study of Religion. I thank Kurtis Schaeffer and Charles Mathewes for their vision in hosting this, the graduate students who helped organize it, and my fellow participants for three weeks of healthy, informative, and challenging dialogue about this thing we call religion.

Within the context of academia, I have also been blessed with the ability to interact with stellar students, at both UF and especially CofC. I thank the many inspiring students I have had in the various religion and ecology classes I've offered over the years; my religion and sustainable agriculture classes; my seminars on radical environmentalism and food security; my spirit of sustainability class; and my ecovillage courses (a big thanks here to NikiAnne Feinberg at SOIL and the rest of the Earthaven crew!).

The other side of my professional background resides within the sustainable agriculture milieu, in the retail environment and more so in the fields of various organic farms. For giving me my first opportunity to pull weeds and harvest good veggies, my thanks to Celeste and George Albers of Green Grocer on Wadmalaw Island; for letting me and Jeanette drive figs all over San Francisco and introducing us to biodynamics, thanks go to Rick and Kristie Knoll; for letting me and Jeanette spend a season living and working on a CSA in New Hampshire, thanks to Tom and Ruth Earle; and for letting me keep my hands dirty while getting my M.Sc., a thanks to Mike and Fu of East Coast Organics in the glorious Pencaitland Hills (and thanks to Robin, Dave, Wee Mike, Jake, and the rest of the work crew).

A number of friends have sustained me throughout the process of getting my doctorate, starting my career as a professor, and writing this book. Here I thank Shannon Horning and clan for the upstairs where much of this writing was originally completed; Andrew for needed brain escapes into video games; the Steeprock crew for good memories; Joel and Ward for playing more hits than misses; Ian (and YuYu) for being a book lover; Reva and Ary for loving good music; Christa and Jason for letting my family escape the humidity and get up to the mountains; Dan, Vinny, and Jack for sustained dialogue on how to be a better human; and Jeremy and Adam for keeping me out of my head and instead running around on the tennis court and paddling up and down Folly Beach whenever the waves picked up. Finally, a smile to my global friends and confidants from Schumacher College's 2000 course Soil, Soul, and Society, and the visionaries who gathered for the 2000 Ecovillage Training at Findhorn in Scotland—I wish we could all rendezvous again to harvest some veggies and share stories.

My mother passed away in 2009 while I was doing research for this book. My love of learning was planted by her when I was very young, and I wish she could have lived to see this finished product. My father planted in me a love of the natural world and farming, and for this I am eternally thankful. We are lucky to live in the same town as most of our family, so I am grateful for many family gatherings with the LeVasseur/Halberda/Hoffman clan over the years. Although not as frequent, Josh, Susanne, and Sophie, I am thankful for your distant support, and Mike, I am thankful for the many talks about politics over the years and for your support of my family as I spent many a night writing.

My final thanks are the most important to give. On one side is the gratitude to the many excellent and inspiring people I was privileged to be able to interview for this book. The hardworking people of Hazon and Koinonia broadly freely gave of their time and knowledge, for which I am grateful because without it this book would not exist. On the other side, my life changed for the better when two gifts entered my life during the time of this project. Aviva Ellianna and Leopold Haviv, I love you both and look forward to more adventures picking blueberries and playing in the garden! Last, my biggest thanks of all are reserved for my wife and best friend, Jeanette Marie Halberda, whose infinite patience and support over many years have allowed me to selfishly pursue farming, research, and writing, on numerous continents.

PREFACE

The postmodern condition lends us to believe that there are vast cultural, political, religious, and philosophical differences extant on our planet. Such a belief is largely fostered by the machinations of modern media: with instant access to news stories and blogs from around the world, it is easy to become mired in a false sense of particularity. However, the reality of this belief is only skin-deep and is a simulacrum (Baudrillard 1994).

Rather, modern *Homo sapiens sapienses* are united in ways that are increasingly brittle. Our base unification is born of a long evolutionary history that requires us to consume approximately 2,500 calories of sunlight energy, coupled with a few liters of clean water, every day. If such quotidian needs are not met, our bodies rapidly deteriorate and die. Such a biophysical reality unites us in our humanity and in our basic needs, as recognized by Abraham Maslow decades ago. This evolutionary and cultural unity has led agrarian essayist Wendell Berry, who figures prominently in the exploration that follows, to quip that there is no such thing as a postagrarian economy.

What Berry means is that if you peel away the detritus of industrial civilization—derivative trading, the stock exchange, global trade agreements, computers and smartphones, SUVs, PTA meetings, the Super Bowl, the Kardashians, civil religion(s), and the rest of industrial and postindustrial society—this truism remains: we need food to survive. For Berry, the true measure of value in a culture rests on that culture's ability to equitably, democratically, and sustainably grow and buy food. To say that there is no such thing as a postagrarian economy means that without sustainably and equitably producing calories as the basis of an economy (taken to mean the management of a household, from local to international levels), then all other additions to household management inevitably collapse.

The global food economy of the early twenty-first century is an industrial one, with a few key commodities such as corn, rice, soya, milk, bananas, apples, sugar, tea, coffee, and wheat providing the basis for everything else.

These commodities are grown at ever larger scales by fewer and fewer farmers and are traded at further and further distances, resulting in more humans living in urban and suburban areas than ever before. On the surface, it appears that the mass of humanity residing in industrialized countries is divorced from food production, yet Berry's observation holds true. For example, the food supply of New York City is approximately three days deep. If trucks were to stop delivering food to this metropolis of over twelve million people, then the true basis of the food economy would quickly strike home. A lack of caloric inputs into the city would lead to widespread chaos and confusion borne of twelve million humans attempting to find sources of calories as the clock approaches seventy-two hours and beyond.

Such a scenario is not farfetched and is why, in terms of systems terminology, our modern industrial food system is becoming increasingly and alarmingly brittle (Meadows 2008). All it takes is a bit of pressure on a key conduit and the system will locally collapse, with very quick ripple effects so that whole regions of the system may follow. With the collapse of food supplies, it is a very quick move to the failure of social systems.[1] Here we need only to analyze the Arab Spring of 2011 and notice the role rising food prices played; combined with access to social media and political unrest, the result was massive social change that will shape the region for years to come. In this scenario, it can be argued the system shift was beneficial for some, given the despotic regimes that ruled the geopolitics of the Middle East and North Africa. However, the long-term instability of the region's food and water supplies suggests such political-ecological stress is only going to be exacerbated. This is seen in one of the subsequent fallouts in Syria, where drought triggered by anthropogenic climate change has displaced numerous farmers into cities. Once in cities, social pressures lead to recruitment into Daesh/ISIS/ISIL, whose terrorist attacks are contributing to destabilizing the region.

Parallel limits are stressing the industrial food system of North America, as the summer of 2013 and again in 2016 witnessed record heat and drought, especially in the Southwest, the agricultural lands of California, and Texas. Furthermore, this drought builds on a record drought from 2012, so new bioclimatic regimes for these regions spawned by anthropogenic climate change have likely been triggered and will only get worse for the rest of settled human occupation in these areas. Climate models predict such weather and suggest that this is the new normal for the heartland breadbasket of the United States. Such weather patterns that are shifting toward being drier and hotter, when coupled with the continued drawdown of the Ogallala Aquifer, itself a nonrenewable resource, suggests that the industrial North American food economy is surprisingly brittle and vulnerable to perturbations. Similar scenarios are developing the world over, from changing monsoon rain patterns in India and Africa, to record droughts in Russia, to record rains in Europe, to massive salinization of soils in Western Australia, to water shortages in the Middle East.

What of East Asia, home to billions? China provides an interesting case study in how calories unite humanity. For the first time in its long history of human settlement, China is importing rice to feed its population of over a billion. This means that China is no longer self-sufficient in its ability to grow this key food staple, and the nation is now a key player in trading for and buying global food commodities. The rice deficit has lead China to begin leasing arable land in developing countries, especially in Africa, where these countries will grow food and ship it to meet China's domestic need for calories. Once the price of oil and gas (for shipping fuel, fertilizers, and tractor fuel) are figured in, combined with the need to police the seas and store the food as it is shipped, the brittleness of the modern food economy is encountered yet again.

Peak oil, the potential for a mass outbreak of a virus in the animals we eat or in the people who grow and ship our food, and climate destabilization add even more pressing biophysical limits to contemporary agriculture. Other limits are evident at smaller levels, such as in the population of bees and other pollinators who are suffering die-offs linked to chemical pesticides. To smaller levels still: a warming atmosphere means warmer soils, so the *Microceleus steenstrupil* cyanobacteria that helps bind the soil food chain in hot climates will soon take over the *Microceleus vaginatus* cyanobacteria that binds soil food webs in colder climates, most likely within fifty years. This shift may well lead to the collapse of agricultural regions in North America that depend on the latter microbe to function, as the soil food webs and trophic chains supported by *M. steenstrupil* do not nearly support the same level of plant life as does *M. vaginatus*. Furthermore, increased carbon loads in the atmosphere may trigger a higher recycling of nutrients in the soil food web, leading to the burn-out and collapse of microbes in both agricultural and forest soils.

Considered altogether, we quickly see that the food that unites us does so in fascinating interlinked culinary, caloric, political, economic, ecological, and cultural loops. These synergistic loops mean that the reality of feeding our bodies on the photosynthesized products of contemporary sunlight unites us in our collective historical and evolutionary past, present, and future.

This book enters into the various loops of this deterministic caloric materialism, and does so through the lens of religion. As will be seen, the modern industrial food economy, as well as the history of the agricultural revolution, is not without its critics. In the pages that follow, I synthesize, share, and investigate various criticisms of the industrial food economy as argued by food activists, scientists, and most important those I am calling "religious agrarians." These critics recognize the inherent brittleness of the industrial food economy and argue that the health of bodies, ecosystems, and democracy are increasingly at risk the longer we continue to use Green Revolution technologies, policies, and worldviews to manage how we grow, market, and consume food.

The early chapters of this book explore the various ideas contained in this last sentence, and this exploration sets the stage for the unique and recent emergence of religious agrarianism and the growing desire to return to (agricultural) place that I argue is occurring in North America. Polls consistently suggest that the secularization thesis is not adequate in understanding the proclivity humans have toward being *Homo religiosus*.[2] It can be argued that this pan-human unity in having beliefs about and practices associated with supernatural agents, broadly (including powerful beings, ancestors with agency, and animate intelligences that can be this-worldly), is a close second to the human need for food as being a trait that unites us all. Such a reality means that the intermixed phenomenon of religion/s and agriculture/s in an industrial food economy becomes fertile ground for exploration.

The book explores this fertile ground, and in so doing, digs up evidence that some religious practitioners in the United States have something profound to say about embodying alternatives to the brittle monocultures of our modern, global, industrialized food economy. Its structure, briefly, is as follows.

In chapter 1, I lay the foundation for key concepts that will accompany us on our journey. These include providing operational definitions of and historical and cultural context for religion, agrarianism, industrial agriculture, and sustainable agriculture. I also weave in insights from key theoretical approaches to the study of religion, including those of lived religion, actor-network theory, and religion and nature theory, culminating in my key finding: recognition of the recent development of a robust religious agrarianism in North American religions. In chapter 2, I introduce the first of my key research communities, Koinonia, and Christian agrarian and New Monastic groups more broadly. This move is echoed in chapter 3, where I introduce Hazon and other Jewish environmental groups more broadly. With this substrate layered in the text, I move to my key chapters that delineate how Koinonia and Hazon are leading exemplars of religious agrarianism. I do this by focusing on three tropes that emerge in my research into religious agrarianism, that when taken together help define it. This is a concern for the local (chapter 4), health (chapter 5), and justice (chapter 6). Chapter 7 brings us to a conclusion, where I harvest possible futures that can help us better understand the vibrant, bourgeoning cross-pollinations between religion, food and farming, and environmental concern.

CHAPTER ONE

SUSTAINABLE RELIGION, SUSTAINABLE ETHICS?

TAKING STOCK: WHY THIS PROJECT?

In this book, I investigate an emerging North American (specifically US) religious agrarianism. I argue that a growing number of religious communities in the United States (with hemispheric and global interconnections[1]) embody many contemporary ecological agrarian practices and values. Three constellations of such values and practices create the foundation on which religious agrarianism is built. These are fidelity to the local; concern for health, including physical, religious/spiritual, societal, economic, and ecological metrics of health; and a deep concern for justice. Although I devote considerable attention to two religious communities—Koinonia Farm in Americus, Georgia, and Hazon, a predominantly national Jewish food group with statewide chapters, including in Atlanta, Georgia, and Gainesville, Florida—where I carried out the majority of my ethnographic research, the trends I discovered in these groups are also found in other religious communities. Given many years participating in and studying alternative food movements, I am confident in predicting that these trends suggest that religious agrarian values are becoming more nuanced and sophisticated. Such development over the past twenty years is leading to exciting new food-based practices in religious communities where such practices are concerned with local places, various types of health, and an abiding concern for justice.

Although other fields of inquiry have either historically or more recently begun to devote sustained attention to the study of agriculture and food, and despite a historic interaction between religion and agriculture, the professional field of religious studies has been slow to come to the table. The result of this foot dragging is that theoretically, the intersection of agrarian and religious beliefs and teachings, coupled with the interaction of their respective values and practices, is an understudied phenomenon of a slowly greening US religious landscape.[2] In this book I take as my starting point this interaction and intersection. This dual focus on agrarianism and religion shapes the following pages, where I analyze and investigate how ecological/environmental agrarianism is combining with and influencing religious beliefs and practices to create an emergent North American religious agrarianism.

RELIGION, THE ENVIRONMENT, AND THE UNITED STATES

Some elements of religions, and especially some religious groups, are undergoing and generating what some scholars call an "ecological reformation." From attending global meetings about climate destabilization to changing to more efficient, eco-friendly light bulbs in places of worship and practice, practitioners of religion are rapidly becoming involved in sustainability initiatives and projects. Such growth appears to be more than a fad and is generated both horizontally from the margins of religions and vertically via institutional creeds and executive decisions. Evidence suggests that there has been tremendous growth in some subsets of various religions in cultivating environmental concern. Much of this growth has occurred over the past ten to fifteen years, as seen for example in June 2015 when Pope Francis released an encyclical about anthropogenic climate change.[3]

Given these recent phenomena, an increasing number of scholars claim that the future landscape of religion will be more decidedly "green." Within this "greening," much is being shaped by post-Darwinian insights about evolution as well as by taking seriously findings from natural and environmental sciences (Taylor 2004).[4]

Such activity has prompted one religious scholar to opine, "No understanding of the environment is adequate without a grasp of the religious life that constitutes the human societies which saturate the natural environment" (Sullivan 2000: xiii). Although partly an essentialist normative claim, this scholar, like many others, recognizes the role "religious life" plays in the natural environment within which humans are embedded.[5] Because humans are embedded with both religion and with metabolizing bodies that need calories to survive,[6] how we grow, harvest, and consume such calories becomes an activity that takes on religious overtones and religious attributes of its own. It is also an activity that can be influenced by religious beliefs, customs, and practices. Therefore, the role of farming, specifically the worldviews behind agricultural technologies and types of farming methods, becomes a topic about which religious environmentalist values, beliefs, and practices are inherently concerned. Religious agrarianism is partly found within this nexus of sustainable food production and religious environmentalism where concerns about the local, health, and justice figure prominently from local to global levels.

As a scholar I accept the growing consensus that future religions (including those in the United States) will most likely be greener, and I couple this acceptance with the recognition that to understand our environments, we must understand our religions. Given that the most recent and emergent US version of agrarianism is largely about the human relationship and interaction with the environment, it is important to understand what religious attributes, emotions, norms, and values attend to this relationship. It is also important to investigate if and how these religious attributes are green, and how these green attributes might be contributing to a larger greening of US religions. It is also important to analyze if and how religious agrarian beliefs and practices are influencing the beliefs and practices of more mainstream religions in the United States.

What is at stake with how we grow, transport, and consume our food? If we as a species are united by the evolutionary dictate of needing calories and living within the laws of thermodynamics, and if we care about issues like justice, equality, biological carrying capacity, climate change, and social power, then with food, everything is at stake. But this daunting list doesn't entirely capture what is at stake. Given tipping points triggered by fossil fuel–based lifestyles, within which industrial agriculture is playing an overly determinant role in anthropogenic climate change and species-wide patterns of land use, our very survival as a species (let alone almost all the others) is now, sadly and scarily enough, at stake.[7] Throw in the tinderbox of failed states, climate refugees, peak oil, epidemics, ocean acidification, nitrogen deposition, and mass starvation and the stakes only get more pronounced. Unfortunately, this is a very real scenario for the immediate future, not to mention decades from now for our children.

I repeat, with this refrain framing my exploration in this book: what is at stake? Everything. These stakes must be recognized in our scholarship and our writing, regardless of field of study.

Since everything is at stake, it is imperative for us as scholars and embedded biosocial creatures to understand what motivates people to produce and consume foods in certain ways. This holds for unsustainable and unjust methods of food production, but also especially for methods of production that can lead to both biological and cultural resilience. The stakes get more dire and important as we enter deeper into the Anthropocene age, or what Paul Gilding calls "the great disruption" (2011). Data from the present and as we look outward a few decades suggest we will pass sobering tipping points, with most caused by human lifestyle choices, where agricultural choices play an overdetermining factor. Such tipping points range from global concentrations of atmospheric gases to the continued loss of soil fertility to decreasing supplies of fresh water to peak oil. All of this and much more will affect the global art of agriculture—and thus the production of religion.[8]

Yet if agriculture is global, why focus on the United States? This is because from the standpoints of biological materialism and the quest for sustainability, the United States currently has about 5 percent of the world's total population of humans, yet as a country consumes 25 percent of the planet's total resources.[9] From the perspectives of environmental justice and the need for humans as biological organismsto live within sustainable limits, this is clearly a problematic figure. The political, moral, and ecological implications suggest that US citizens have a profound duty to ameliorate such discrepancies.[10] In claiming this, I must be cautious of engaging in neocolonialist discourses of ecology, so we must clearly recognize that the imperial eco-footprint of the contemporary United States and its unique history of environmental movements is embedded within and constructed by issues of power, justice, complexity, and essentialism (Roos and Hunt 2010; DeLoughrey and Handley 2011).[11] Such baggage may also accrue to narratives about sustainable agriculture and religion in a US context, and this reality must be remembered when talking about issues of food and cultural fluidity.

That said, there is an equally articulate, sustained, and active history and contemporary environmental movement, or rather movements, within the United States.[12] The books and articles devoted to this aspect of the US gestalt are legion and growing. Also growing is the consciousness within the populace of the states about environmental issues, ranging from climate destabilization to sustainable agriculture (van Wormer et al. 2007). It is foolhardy for scholars of North American/US religion, religion broadly, and the environmental humanities, as well as activists and policy makers to disregard this history. It is as foolhardy for scholars to disregard the growing impact such environmentalist beliefs and values play in US religious production: past, present, and especially the foreseeable future.

Because religions are in large part reified social constructs that have fluid, changing, and contested concepts of what qualifies as sacred, holy, and/or legally obligatory relationships and duties, it becomes the job of religion scholars to study how and why these concepts and categories are changing. Given what is at stake, scholars must pay close attention to how religious concepts and categories relate to nature and environmental issues.[13] One area of human/nature relations is the human (and thus religious) relationship to food and how food is produced and consumed. Given the material environmental dimension of food production; the value-laden social, political, and ecological goals and normative claims of agrarianism (and of industrial agriculture; Sanford 2012); and how these relate to sustainable food issues (which are part of a larger sustainability milieu) that some religions/religious bodies are becoming cognizant of, then the study of this emerging environmental dimension of US religions becomes an area of concern and study that scholars are beholden to engage. I hope this book helps us better understand the values and norms that entail to this environmental dimension of US religions and to the sustainable farming practices in the United States. Furthermore, I hope it adds to the growing body of work that clearly spells out what is at stake with our food choices given the social and environmental impacts of food production, distribution, and consumption.

A further reason this book chooses to study religions in the United States is because of the lengthy historical role US religions have played in politics. With this dynamic and rich past, parts of this book explore this legacy of US religious history, helping shed light on religio-political interactions at the intersection of agriculture and sustainability.

Last, compared to other Western nations, religious identification and belief still constitute a large part of the identity of the vast majority of the US public. Various polls report that nine out of ten US citizens believe in God or a higher power, and others suggest that over 70 percent identify with some variety of Christianity. Coupled with the aforementioned polls about environmental values and concerns within the US public, it behooves scholars to investigate the interaction of religious belief and environmental behavior in the United States. This does not necessarily mean there is a causation between religious belief and environmental practice, but that such study should be undertaken to help the community of scholars understand if such links exist and, if so, how pervasive and effective they actually are.[14]

For example, religion and nature scholar Anna Peterson comments that "changes in values lead[ing] to changes in behavior [is] an assumption supported by little if any empirical and historical evidence. The paradox of modern environmentalism is that while pro-environmental values have become mainstream in the US and many other parts of the world, anti-environmental practices continue to escalate" (2006: 376). Despite this gap in knowledge, she continues, "Religion is indeed the way many people think about important moral and intellectual issues, and an expansion of environmental ethics to address religious traditions and ideologies is an important and necessary step" (ibid.: 378)—if true, then this is a step to which religious scholars must pay attention. Of course, such debates about religious belief and action toward the natural environment date back to Lynn White Jr.'s (1967) epochal article that blamed Western Christian beliefs, and thus actions based on those beliefs, for the world's environmental problems. By focusing especially on the iterative and reinforcing interaction of religious values and agrarian praxis, an understanding of religious agrarianism adds to this growing body of knowledge about religious belief and practice regarding environmentally centered behaviors. If a key goal is to understand this iterative and reinforcing interaction, then it is important that I use appropriate theoretical and methodological tools that help the process. I briefly outline these choices in the next three sections.

THEORY AND METHOD: LIVED RELIGION

Readers will notice that throughout the book I adopt various scholarly methods and analytical views. Of these, the two key poles are a sociological point of view (how do religious agrarians act as a group, how do they organize, and how do they make sense of their standing in a plural society?) and an ethical point of view (what specific views about appropriate treatment of the natural world do religious agrarians hold, how do they embody these, and how are they informed by religion?). To dance between these poles, I weave together ethnographic findings,[15] historical sleuthing, and constructive ethical building. This syncretic approach mirrors the interdisciplinary raison d'être of religious studies and recognizes that there are multiple approaches to better making sense of real-world data (Strausberg and Engler 2014). Because religious and environmental values are so central to an understanding of religion and nature issues—and sustainable farming in particular—it is important to pay attention to how these arise, are shaped, and are formed. I have found that a "lived religion" approach best makes sense of how religious subjects shape religious beliefs and values, and in turn are shaped by the larger networks in which they reside.[16] Even more important, this approach allows scholars to see how nonspecialists shape their own religious worlds, at times in concert with and at times at odds with the larger top-down traditions in which they are participants.

This is because a lived religion perspective takes the symbolic, material, ideational, affective, and emotional worlds that lay religious practitioners create and privileges these worlds as being worthy of study. Rather than privileging the religious teachings, theologies, and institutions of elite experts, lived religion scholars head to the street, literally and figuratively, to see how the average person within a religion is creating a meaningful religious world that works and makes sense for them and their needs.

Robert Orsi's *The Madonna of 115th Street* (2002) remains the standard-bearer of lived religion scholarship.[17] Orsi's introduction to the updated version contains some of the key theoretical insights and methodologies of a lived religion approach, most importantly recognizing that

> The study of lived religion situates all religious creativity within culture and approaches all religion as lived experience. . . . Rethinking religion as a form of cultural work, the study of lived religion directs attention to institutions *and* persons, texts *and* rituals, practice *and* theology, things *and* ideas—all as media of making and unmaking worlds. . . . Religion approached this way is set amid the ordinary concerns of life as these are structured at various moments in history and in different cultures, at the junctures of self and culture, family and social world, and on those occasions when the religious imagination . . . takes hold of the world [and] as it is taken hold of by the world. (2002: xix–xx)

Many of Orsi's claims and insights are reflected in my approach to studying religion and thus to my interpretation of religion and agriculture. For example, my ethnographic research with members of Hazon/CSI and Koinonia only makes sense situationally, including within larger debates and ideologies about agriculture, politics, and sustainability. These concerns are reflected in the practices and values, religious and secular, of these Jewish and Christian religious agrarians as they attempt to live out their religious agricultural sensibilities. Their attempts highlights what Orsi calls "religion-in-action," which is "religion-in-relationships between people, between the way the world is and the way people imagine it to be" (Orsi 2002: xx). It is essential to recognize, at least for religious agrarians, that this also includes religion-in-action with the larger biological world, of which agriculture is one important aspect. Thus, joining a community-supported agriculture (CSA) farm, planting a row of nitrogen-fixing legumes, humanely raising a pig—all have the potential to become embodied religious acts within a lived religious agrarian worldview and help signal the reemergence of place within the religious agrarian milieu.

Moreover, religion-in-action is always transforming, is influenced by and representative of relations of power and gender, affects culture at large, and is also affected by culture at large. This is why for Orsi "the study of lived religion vividly opens out the tremendous creativity of religious practice and imaginings as it uncovers the limits of them" (2002: xxiv), leading to what he calls a "hermeneutics of hybridity" (1997: 11). I argue that the lived religious landscape of the contemporary United States is generating tremendous hybridity, creativity, and imaginings in regard to a plethora of environmental issues—and at the same time, this creativity has its limits when the harder work of translating values into practice becomes evident. Agricultural landscapes thus become a central focal point where religious environmental imaginings and practices are lived out and transformed, bringing with it ripple effects throughout the larger US cultural, religious, environmental, and political landscapes.[18]

In concluding this section on lived religion, I make clear that my own interpretation of the religious agrarians I interacted with and the research data I generated is thus influenced by my training in religion and nature and US religious history; by my personal experience living and working on organic farms and in natural food stores in South Carolina, California, Washington, New Hampshire, and Scotland; and by being exposed to sustainability subcultures via this history and studying ecovillages, soil and society, and ecophilosophy in Scotland, England, and Australia (and by being a Willing Workers on Organic Farms volunteer in Greece). My interpretation, and indeed this whole book, is not to be reified but should be read as a sighting from a site (see Tweed 2006: 13)—including the physical sites (congregational buildings, monastic campus, conference centers, telephones and computer, urban cityscapes and rural landscapes, shopping centers and grocery stores, and farmlands) where my research was undertaken.[19]

THEORY AND METHOD: NETWORK THEORY

Because food items, sustainability discourses, and religious institutions are now global in their flows, I use a network approach to studying religion and nature issues. This approach works in tandem with a lived religion approach, as the network approach is "attentive to ways in which local, grassroots, official, national, and transnational actors continuously and creatively construct, transgress, and appropriate the boundaries between specific religious and non-religious practices and discourses. These multiple situated perspectives (which often lead to contested canons, traditions, and orthodoxies), in interplay with the researcher's own personality, determine religious studies' proper subject matter" (Vasquez 2005: 237).[20] I echo this statement and urge religious studies to actively take as its subject matter biological flows of food (and other material objects) and the religious environmentalist values, discourses, and practices that these flows help create.[21]

Vasquez builds on his claim, writing, "Within and through networks, actors carve out spaces to dwell, itineraries, and everyday routines, drawing from religious symbols and tropes to reflect on and orient their own praxis and to 'sacralize' nature and build environments. Networks also embody and produce moral geographies" (2008: 168–69).

Given this insight, what I am calling a "lived-network" approach allows us to make sense of the ways religiously and environmentally concerned US citizens are sacralizing nature, building agricultural environments, and constructing and participating in sustainable agriculture networks that are local, national, and international. Importantly, for political, ethical, economic, and biological reasons, these networks also help produce moral geographies. As David Chidester and Edward Linenthal write, "the 'pivoting of the sacred' that occurs through the work of ritualization and interpretation allows virtually any place to become sacred . . .sacred meaning and significance, holy awe and desire, can coalesce in any place that becomes, even if only temporarily, a site for intensive interpretation" (1995: 14). If they are correct, and if we take Vasquez, Orsi, Tweed, and others seriously, then the sustainable agricultural landscapes, farmers' markets, and religious networks that support and advocate for these "sacred" sites become a product of US religion that needs to be studied. They also become sites where the power-laden, religion-in-action, lived praxis of making place sacred occurs.

THEORY AND METHOD: GROUNDED THEORY

As this is a study of lived religion, I use ethnographic methods that help capture the complexity and vibrancy of religion-in-action that emerges via religious agrarians navigating and shaping complex, interrelated networks. These networks reflect and construct environmental, religious, and agricultural worldviews and material landscapes. My research is therefore guided by the following sentiment expressed by Emerson et al., who write, "The ethnographer's central purpose is to describe a social world and its people" (1995: 68). This means that during the fieldwork for this book, I was motivated by the following research questions. Who are the key actors in the social situations that I am witnessing and in which I am involved? What concerns are expressed or glossed over in the various interactions of my interlocutors in the (religious) social world they shape and are shaped by? What interactions with the natural world are occurring, and why? Last, what worldviews and ethical and affective and emotional concerns help shape the varied human behaviors toward the natural world in the social situations I observe?

I am also motivated in my ethnography and fieldwork by the following claim from Borneman and Hammoudi, who write that the "fieldwork experience . . . provides an opening to dilemmas in the contemporary world" (2009: 18). Sustainable agriculture issues are indeed a mounting dilemma of the contemporary world, especially given what I outlined earlier about what is at stake when it comes to food production. The social world of religious agrarians is able to offer insights into how some members of various religions in the United States are responding to this dilemma of how to sustainably meet our caloric needs while contributing to social and biological resiliency.

Besides undertaking ethnographic fieldwork based on a variety of means of recruiting participants and that lead to semi-structured interviews and biographical sketches (see the Appendix), I used grounded theory methodology to help learn about the social, religious, and ethical worlds of religious agrarians. Grounded theory is a subset of qualitative research and is an inductive approach to generating knowledge about social worlds used to generate theory by interpretation of collected data. Qualitative research investigates "words rather than numbers [and has] an epistemological position described as interpretivist, meaning that, in contrast to the adoption of a natural scientific model in quantitative research, the stress is on the understanding of the social world through an examination of the interpretation of that world by its participants" (Bryman 2001: 264).

Grounded theory is the "discovery of theory from data" (Glaser and Strauss 1967: 1–3), versus discovering theory from a priori assumptions. Glaser and Strauss call grounded theory "theory as process" (ibid.: 32), meaning the social world described by it is never claimed to be complete and does not offer one-size-fits-all claims. Rather, grounded theory is open ended and as more research is carried out over time, with reengagement with those being studied and observed, theory is built on, changed, and developed. Such an approach is consistent with the lived-network theoretical lens I adopt for my research—because the values, practices, beliefs, and networks of those I researched are fluid, grounded theory provided me with methodological tools to capture this fluidity. It also allowed me to develop a more nuanced understanding of the various discourses, rationales, and relationships that enter into shaping the value-laden and fluid religious, biological, and social worlds of religious agrarians.

By using grounded theory, I developed my own trope and category of "religious agrarianism," which I more thoroughly explain later in this chapter. Next I give a brief history and overview of the two key communities I engaged. This is followed by a brief summary regarding the importance of religious values and then a turn to exploring sustainable agriculture. The chapter concludes with an exploration of food, agriculture, and agrarianism, focusing especially on the latter's development within the United States.

RELIGIOUS AGRARIAN COMMUNITIES

Within the context of US agrarianism to be outlined here, and especially US religious history, I have chosen two faith communities to serve as case study exemplars of religious agrarianism. Chapters 2 and 3 give greater detail about these communities so I only give a brief introduction to each here.

One is a historically Protestant lay-monastic community called Koinonia Farm. Koinonia is a 501c3 profit-sharing intentional community that is located on an approximately 600-acre campus. The community consists of about twenty full-time members and others who live on site as interns or as guests in residence to see if they would like to become members. Clarence Jordan (1912–1969) founded Koinonia outside of Americus, Georgia, in 1942 when Jim Crow racism was still rampant throughout the region. Dallas Lee, biographer of Koinonia, writes that Jordan "was a dirt-farming aristocrat, a good 'ole Georgia country boy with a doctor's degree, a teacher with manure on his boots, a scholar with working clothes on his mind" (1971: 1). Jordan began Koinonia as an experiment in interracial living, inspired largely by his doctorate in Greek New Testament from the Southern Baptist Theological Seminary. He also earned a degree at the Georgia State College of Agriculture at the University of Georgia, Athens.

Jordan's vision was to unite the races in a voluntary life of simplicity—in effect, a religious experiment (drawing especially on the book of Acts[22]) based on shared farming responsibilities. Jordan and his wife named the community Koinonia, from the Greek term *koinonia* being translated into "fellowship" and "communion." In the summer of 2009, I spent two weeks living in residence at Koinonia, conducting interviews and participating in farm duties with the members and interns of the community. Along with repeated follow-up phone and e-mail interviews and return trips, including for a 2010 permaculture training, this experience is the basis for my analysis of Koinonia's approach to what I argue is partly a continued experiment in intentional Christian agrarian living.

The other religious community that features prominently herein is Congregation Shearith Israel (CSI), an egalitarian Conservative Jewish synagogue in Atlanta, Georgia. CSI is partnered with Hazon, a recently formed (circa 2000) US progressive Jewish food group with key offices in New York City and San Francisco. Hazon works with various Jewish temples and synagogues around the nation and helps them find local community-supported agriculture (CSA) farms that synagogue members can join. My research for this case study exemplar consists of repeated visits to CSI, visits to Riverview Farms in northwest Georgia (the CSA that CSI partners with), interviews with the rabbi and various members of the synagogue and those responsible for organizing the CSA-CSI partnership, and participation in two of Hazon's annual food conferences.[23] I also interviewed members of B'nai Israel, a synagogue in Gainesville, Florida, that began a partnership with Hazon in late 2009. Together, this fieldwork allows me to speak about Hazon broadly and the two Jewish communities in particular.

Jewish and Christian communities are the two predominant types of religious groups in the United States.[24] This especially holds in terms of membership for Christianity and in terms of the social and financial networking maintained by adherents of both religions. Meanwhile, followers of certain branches in these religions are extremely active in lobbying the US government, and this political capital is a potential resource for religiously motivated political concerns pertaining to agrarianism. This religion-government interaction is built on a unique history where Judaism and Christianity (and especially appeals to the Hebrew Bible and New Testament made by citizens within these religions) have contributed more than any other religions found in the United States in advocating for progressive social change. My research shows that parts of this legacy are now advocating for progressive agrarian and environmental change via modern religiously based agrarian and religious environmentalist values.

Yet this agrarianism is similar across Jewish and Christian religious identities, while at the same time being distinct for members of both. Understanding such similarities and differences will help scholars understand emerging concerns and motivations that attend to US religions and how these religions interact with and are shaped by environmental issues. However, it should be made clear that I do not speak as an insider for either Hazon or Koinonia, let alone the many other nuanced varieties of religious agrarianism in US Christianity, Judaism, and other emerging forms of religious agrarianism. Clearly there are other Christian and Jewish religious agrarian groups in North America, who will be similar and different to those in Koinonia and Hazon, and further studies are needed to help tease out such data.

Finally, despite the migration of ideas, bodies, seeds, and pollutants throughout the United States, there are regional differences when it comes to culture, history, and environment (Cowdrey 1996), and this is also true for religious agrarianism.[25] This book uncovers and lightly sprinkles in some of this regional flavor.

WHY RELIGIOUS VALUES?

We will see shortly how any kind of agriculture involves religious, ethical, normative, affective, and value-laden claims that reflect a worldview. For religious agrarians, the values that permeate their view inspire and motivate them to actively participate in sustainable farming practices. This is because religion structures people's lives in many different ways. A religion may be operationally defined as a contested yet systematic set of narratives, values, practices, teachings, doctrines, experiences, social grouping, material items, and myths that are maintained and passed on by charismatic figures and institutions. The latter typically put forth a cosmology that is tethered to a belief in some sort of supernatural being/s, agent/s, and/or realm.

Religious beliefs and identities offer sanctuary in times of need, inspiration in times of hope, and a sense of grounding in today's wayward, fractured, sped-up world. Religion also has many functions in peoples' lives. Belief and membership in a religious community can provide stability and a sense of purpose. The myths, narratives, and practices within a religion also provide rituals and influence practices like pilgrimage and prayer. They also offer a code of ethics to live by and an annual calendar of activities in which a person can participate. Religion (or rather, religions) also help people generate and codify values so they can attempt to live a life in line with the ideals and teachings of their faith and with their own personal ethics (which for many are largely shaped by the religion to which they belong or were raised in). Religious values also help those in the United States navigate a contentious political system and a diverse religious landscape (Eck 2001).

Religion is also importantly about difference, negotiation, and conflict. This can occur over competing conceptions of the sacred, beliefs and doctrines about where and how the sacred manifests, to competing values and ethics, to differing interpretations of felt experiences. According to Ninian Smart (1998), such conflict results in the formation of subtraditions within larger traditions. This conflict leads to the development of new religions or bricolage between existing subtraditions, resulting in very real power-laden issues of authenticity (Martin 2012; see especially chapter 7).

This history of religious conflict is also true in regard to conflicts over value-laden food choices. Here choices include those that explicitly support sustainable agriculture, as well as those that by default gloss over any understanding of how food is grown so that this distancing is itself a subconscious (or even conscious) normative position. Given that "religion is the most volatile constituent of culture" (Prothero 2007: 5), it is important to develop religious literacy. One area where this literacy is lacking is what groups within a religion have had to say about food and agriculture. This refers to the past and the present. Religious values around food choices are especially important to understand given the current ecocrisis, including specifically the values that underlie decision-making regimes related to food production and farming methods. This book is one corrective offered from the field of religious studies that can help us better understand this interaction between religious values and food/farming choices.

Furthermore, in a highly religious country such as the United States, religion and religious values play central roles in many domains: political, ethical, economic, institutional, and increasingly environmental.[26] Because values are central to discourses about sustainability and thus sustainable agriculture, it is important to be clear on what we mean when we talk about values.[27] It is equally important to be clear about what we mean by "religious values" and how these relate to the values of sustainable agriculture.

Many values held by US citizens are religiously based and/or affected, such that "The clue to American values, including religious values insofar as they can be separated, must . . . be sought in the American revolutionary tradition. The fundamental elements in this libertarian social system were . . . the security of property [and] a body of thought whose most relevant origins can be traced to the leading Puritan theologians and social thinkers" (Van Allen 1978: 20). Max Weber's (2003) thesis on Protestantism and the rise of market capitalism in the United States is also relevant here. If Weber and Van Allen are correct, then one of the underlying bedrock values of the United States (secular and religious) is a right to private property and the right to do with this property as the owner pleases.

Some environmentalists criticize this claim, arguing that market forces (especially subsidies) only reinforce the unsustainable abuse of US ecosystems. This includes the "productionist" industrial agriculture model, where corporate behemoths are fiscally rewarded by the right of farmers to do what they please on their privately owned lands.[28] Many who make these arguments, even if they acknowledge that the US Department of Agriculture and other governmental regulatory agencies adequately monitor industrial farming practices (including the use of toxic chemicals), claim that nonetheless the government is beholden to agro-corporate interests and that these regulatory mechanisms are unsatisfactory and weakly enforced. This debate is central to the sustainable/industrial divide, and it exemplifies the politically contested landscape of agriculture and its underlying values in the modern United States.

Such a historical understanding of values (and rights) has led to an extreme individualism that has become the de facto dominant value of modern North America (Bellah et al. 1985: 22). Many today add that this privileging of the individual does not lead to the development of sustainable environmental values or environmental success—in fact, one of the insights of ecology is that all life forms (and all systems) are interconnected. When one life form becomes privileged in nature, it tends to eventually encounter limits that lead to its demise or systemic processes bring it into balance with the rest of the system. This argument holds that privileging individual, rational economic man [*sic*], and the value system based on this view of humanity, is one of the key contributors of the current ecocrisis.[29]

The communities I study provide an alternative understanding of what it is to be human. This is one reason we should study religious institutions in general: religious institutions are sites of value creation and maintenance, including possibly deviant values, especially compared to the extreme individualism of US society.[30] Scholars can also investigate if the values held in religious communities temper the individualism of modern North America or if they reinforce this individualism. One clear example of this is seen in the "health and wealth gospel" of some Pentecostal churches.

However, there often exists a gap between the ideal of a professed value and the reality of putting it into practice. This holds true for religious ethics and values, ranging from religious values that guide a marriage to those that guide a parent–child relationship. Such a gap is especially true for environmental values, which may be restrained from being put into practice by competing internal interests, values, and desires and especially by systemic limits within our larger political and consumer systems.[31]

Regarding specific values and practices related to food, Anna Peterson writes, "Consumption and food issues underline the gap between what we say we care about and how we act. In no other sphere of life, perhaps, do people contradict their own values so regularly" (2009: 108). This quote highlights the complex, contested internal and external nature of trying to align religious and environmental food values with practices that enable these values to be realized. This is made even more difficult when we recognize that "Until recently, Western virtue ethics has never recognized a nature-focused virtue. . . . Unfortunately . . . a capitalist, consumer culture such as that of the United States . . . presents a variety of obstacles to successful inculcation of any new environmental virtues" (Fairbanks 2010: 80).

However, values do not adhere only to individuals. A group can also express them, and they are equally shaped and challenged by other groups and/or membership within a group. As can be imagined, this process is loaded with power struggles and politics. It is also influenced by gender, levels of education and income, and prestige awarded to certain members. This process of value emergence occurs in groups (kin, sport, political, institutional, religious) and all group members participate in it. Because putting sustainable agricultural values and their varied motivations into practice is so difficult, it might be easier to embody and practice them in a group setting. This group setting can be a food co-op, a CSA farm, a farmers' market, or an intentional religious community that practices sustainable agriculture. The various religious agrarians highlighted herein are all members of some type of these group settings. Even more central to my argument is that many join precisely because these group settings enable them to practice their sustainable agriculture, religious environmentalist, and religious agrarian values.

A corollary assumption follows: most people are used to the values of cheap and abundant food. These values are inculcated in the United States by the ubiquity of fast food chains and omnipresent food advertising and marketing. These values are a default result of the cost of food remaining relatively constant despite inflation in other areas of the economy. Here we can think of Bellah's laments about hyperindividualism in the United States—the right of individual consumers to buy food shipped around the world for a low price is a pinnacle of individual choice, one that externalizes the ecological and social footprints that adhere to this value-laden choice.

Data suggest that there is a shift occurring in the United States when it comes to values related to food choices and to food choices themselves. This shift is part the result of the interaction of politics and values, economic incentives, and people's social standing and values. Such interaction is already under way in religious agrarian communities. Such a vanguard position means religious agrarian groups have the potential to shape the values of individuals and, through those individuals and their involvement with larger society (through economic transactions, volunteering, and political lobbying), the values of society. This ripple effect will shape agricultural farmlands, soils, waters, and animals.

Attitudes within religious agrarian groups coalesce around a "value orientation" (Van Deth and Scarbrough 1995: 22) so that religious agrarian "values are distinctive not only because they are desirabilities, but, more importantly . . . they are *desirabilities in matters of action*" (ibid.: 29-30; emphasis in original). As will be seen, the people in the South and around the country with whom I interacted translate their desires (or more accurately, their value orientation) into practice by engaging in the actions and habits of sustainable agriculture. By putting matters of values into action they are creating what I call religious agrarianism. The value orientation of religious agrarians is explicitly grounded in religious environmentalist concerns. These concerns affect the values and thus actions of religious agrarians.

In terms of food choices and values related to them, Tina Huey argues, "Perhaps the most visible topic of contention in the globalization of capital, aside from the labor practices of branded clothing manufacturers, is food" (2005: 125). Food (and thus farming) is again seen as a globally contested trope—one imbued with values and with a concomitant growing religious discourse about specific sustainable food values for a growing number of practitioners.[32] Her study concludes by arguing that "local solutions are not privileged in the debate about food in the North" (ibid.: 135). I challenge this conclusion. Rather, I argue that this perceived lack of privileging local solutions is changing. Local solutions are becoming privileged. I argue that religious values are playing a role in this process as religious agrarians embody and bring to the fore local, place-based solutions that are becoming more central to the debate, especially to the solutions proffered, of food issues. Further insights into values and how these relate to food choices are seen in a study by E. P. Koster, who uses food choice methodology to analyze "the central question in food choice research: '*Why* does *who* eat *what, when,* and *where*?'" (2009: 70; emphasis in original). Koster's research highlights the many variables and factors that contribute to what we eat, where, when, how, and why, from biology, physiology, and decision psychology to marketing, economics, and learning psychology (ibid.: 70–71). Religion and religious values are two of these variables, and important ones. Koster posits that "with respect to food habit formation and change much depends also on so-called sensitive periods in life. Thus, most of our basic food habits are formed in infancy and early childhood and these are hardest to change, but periods like late adolescence when people start living in pairs and form their own 'traditions' and other major re-orientations in life such as divorce or retirement are good moments for changes in food habits as well" (ibid.: 75). Tellingly, there is no mention of religion (although there is mention of "traditions"), whether religious ethics, taboos, customs, or rites of passage. Yet religious communities to which people belong tend to influence their members throughout their lives. This includes when people enter into a new community or when a community attempts to embody new values, all of which have the potential to affect food habits.

While for Koster, "food choice is a learned behavior" (2009: 75), it is heavily influenced by values and social peer groups. The influence of peer groups is a determining factor for religious agrarians throughout North America who support sustainable agriculture practices precisely because they do want to influence the diets, health, lifestyles, and values of their children and those in their respective religious communities. Hazon specifically wants to change the food practices of North American Jews, and Koinonia hopes to affect the food choices of those who visit their campus. Both groups are united in that they would like the agricultural choices of governmental policies to change as well.

Finally, regarding environmental values more broadly, including concern about sustainability, polls suggest that a vast majority of citizens are concerned with environmental issues ranging from climate destabilization to species extinction to pollution. Kempton et al. point out that this concern "is not strongly related to social elites," (1995: 7) but permeates many sectors of US society.[33] Furthermore, Wall (1995) and Taylor (1995) provide arguments which claim that people tend to be motivated to care about the environment when they perceive risks to their health or the health of those they love.[34] Such an anthropocentric value system need not be at odds with environmental protection, but suggests that many people are motivated to care for the environment because of values and genuine concerns related to personal health, safety, and aesthetics. The research summarized in this book challenges and supports this argument, as will be explained.

What is important to note is that the aforementioned studies avoid discussing religion directly. In comparison, I argue that we cannot understand agricultural systems and practices without attending directly to religion. These studies nonetheless tell us important things about general food values. What they suggest is that food choice decisions are multifaceted, and peer groups and social values play a key role in such decisions. Furthermore, any understanding of food and farming in the United States and of environmental problems and movements generally must address religious values. Religious agrarianism provides an excellent case study for exploring the intimate biocultural connections between and mutual shaping of religion (values, practices, institutions, politics, and teachings) and the environment.

WHY STUDY FOOD?

Mammalian bodies need calories to survive. Humans consume and imbibe liquid and food items, which are then metabolized so that our cells, muscles, tissues, organs, and blood receive nourishment so we can move, work, play, engage in religious practices, reproduce, think, and contribute to society. Given that eating food is a central act in which humans participate, coupled with another pan-human universal, religion (Boyer 2001; Atran 2002; King 2007; Counihan and Van Esterik 2008), as scholars we need to dedicate more attention to the interplay between the two.

In regards to US religion—in particular its historical relation with food—Daniel Sack writes, "Food carries a moral value in America. In this culture, a particular foodstuff is not only good or bad for your body but also can be good or bad for your soul. Whether based on popular culture or on scientific studies, personal food choice becomes an ethical calculation" (2000: 185). A key argument of this book is that food indeed carries a moral value for religious agrarians. Even more food choices carry explicit religious values and reflect ethical deliberations that are environmental in calculation.

This religious-ethical turn means that how we get our calories becomes an exercise in value reasoning and deliberation, with menus and plates becoming loci of meaning making. As Jeremy Benstein, founder of the Heschel Center for Environmental Learning, states, "few acts are as imbued with as much religious symbolism and stricture as eating" (2006: 150). The meaning making of food choices has even led one advocate of sustainable farming to claim that "Food may be the primary arena where we humans sort out our values" (Holthaus 2009: 258).

Debates about religion-values-ethics-practices and how these relate to food and the environment provides a major context for this book. This is because it is in the realm of the meaning-making, value-laden power of food choices, and the role religion plays in these choices, that religious agrarianism in large part manifests. My findings are clear in suggesting that religious-based values and ethics do indeed motivate people to put into practice sustainable, environmentally friendly behaviors (itself a problematic concept). For religious agrarianism, food is an important marker of value, identity, and custom as it has been in almost all human cultures. Furthermore, food is a point of contention in our current US consumer culture and in our socially constructed religious systems. Although food can be fiber that is sweet, bitter, salty, and sour, it can also be a polysemous material item that becomes a site of contested religious meaning making (Chidester and Linenthal 1995; McDannell 1995; Tweed 1997). Finally, via food aid and governmental subsidies, food is a contested political object, and via genetic engineering, it is a contested scientific object (Kent 2005; Bello 2009; Nützenadel and Trentmann 2008). Given the production of religious, ethnic, political, and material meanings that entail to food and our physical need to consume it, it makes sense to use food as a lens with which to analyze issues related to both religion and nature and US religion.

While studies like Sack's (2000) look at the role of food in US religious history, or those like Shortridge and Shortridge (1998) look at food in a religious context and how it acts as a boundary or marker of ethnicity, my study looks at food in a US context of heightening environmental awareness. Food is not only an environmental issue but also a religious environmental concern.

WHY STUDY FARMING?

The evidence for this turn toward food as a social and environmental issue in contemporary US religions becomes clearer once we look at the material actions, implications, and goals of both conventional (i.e., industrial) and sustainable farming. To help articulate these goals, I give a brief history of what the term *industrial agriculture/farming* means and then contrast this with sustainable agriculture/farming. This history helps highlight how farming and growing food becomes a central ingredient in understanding religion-environment-food issues in the modern United States.

INDUSTRIAL AGRICULTURE/FARMING

Over human evolutionary history, we have adopted various approaches to procuring food calories, including foraging, hunting and gathering, nomadicism, fishing, pastoralism, swidden agriculture, and settled industrial agriculture. Most people in the United States obtain their food from the modern industrial farm. Such a farm is based on a monoculture of a commodity crop—typically corn, cotton, or soybean—planted at scales in the hundreds to thousands of acres.

This type of farming is thoroughly mechanized, relying on large tractors, combines, and geographic information systems (GIS) where crops are planted and then the fields are left barren and fallow over the winter. Fields are graded to be flat, with fencerows and habitat for other organisms removed to ease in helping plant and harvest. Commodity crops are grown with various agrochemicals, ranging from insecticides to herbicides to fungicides to artificial fertilizers. The harvested commodities are shipped to various distribution centers where they are packaged or converted to other commodities. As Eric Freyfogle writes,

> The land manager begins the new year by killing every living thing on the land except microscopic organisms. A single species of life is then introduced, perhaps a bioengineered wonder quite different from any wild plant . . . The plant's seed or fruit has been engineered for appearance, 'shipability,' and ease of harvesting. There is little concern about nutrition or ecological effects. . . . If a human hand touches the resulting produce, it is likely that of a low-paid worker on an assembly line that washes, sorts, and packs. . . . The field is a cultural emblem of who we are and of how we understand our place. (2007: 8)

The purchase and consumption of the vast majority of farmed (and processed) industrial food items grown by the methods described here take place in supermarkets that dot the landscape. Almost one quarter of all US citizens daily consume at least one meal from a fast food restaurant; such consumption makes these restaurants some of the largest purchasers of industrially farmed food. The food items purchased in these restaurants and supermarkets are grown domestically and internationally, as the modern industrial agricultural system is now global in its reach and impact.

It is important to note that all of the foregoing is a recent phenomenon, made available because of the Green Revolution, access to fossil fuels, and technologies of refrigeration and shipping that allow for food items to be shipped at great distances. Our ability to understand, measure, and longitudinally track the impacts of various farming regimes is also a recent phenomenon made possible by satellite technologies and more sophisticated instruments that measure soil and plant health. Overall, the ability to purchase and consume the wide variety of food items available on supermarket shelves and on restaurant menus is largely the result of cheap, abundant fossil fuel energy; refrigeration and shipping technologies; the invention of new, larger, and more powerful tractors and combines; a centralized government and energy grid; fossil fuel–based weapons developed during World War II that became the basis of modern synthetic agricultural chemicals; the development of the land grant university system; international capital, trade agreements, and transport; and corporate and university-sponsored research into improving and creating new hybrid strains of seeds, including especially the work of Norman Borlaug (Kunstler 2005; Pollan 2006; Conkin 2008; Carton 2012). The development of hybrid seeds, along with developments in new technologies for irrigation, refrigeration, and distribution of the products grown by these seeds (and policies and subsidies that favored the adoption and use of these technologies), is commonly referred to as the Green Revolution (Leaf 2012). Over the past seventy years, this revolution has allowed *Homo sapiens* to bring the most land in the history of agriculture into cultivation, has created an abundance and surplus of calories never before seen, and has helped contribute to exponential population growth.

This revolution has not been without sustained, articulate, and impassioned critiques. For example, Rachel Carson (1962), Wendell Berry (1977), Andrew Kimbrell (2002), Vandana Shiva (2000), and Raj Patel (2007) have made strong criticisms of the industrial agricultural system that emerged out of the Green Revolution. Along with others (Jansen and Vellema 2004; Clapp and Fuchs 2009), they lambast the centralized corporate and governmental control of such farming, claim there are deleterious environmental impacts that the rampant use of agrochemicals has on birds and other flora and fauna of farming rivers and landscapes, and argue that transnational agribusiness corporations undertake biopiracy and colonization of food items, seeds, and agricultural lands. The gestalt of emerging critiques against industrial farming includes arguments against food aid policies, global trade agreements (targeting especially the World Trade Organization), criticisms of genetically modified organisms (GMOs), and criticisms of the modern supermarket system that have developed around selling cheap food (in terms of cost and nutrition) in carefully controlled retail environments.[35] These and other scholars and activists further argue that industrial farming uses toxic chemicals that poison our bodies and landscapes and lead to the salinization of soil, rapidly draws down aquifers, and contributes to climate destabilization. A shared concern is held about the effects that mechanized, corporate-controlled farming in the United States has had on rural areas, so that today less than 2 percent of all US citizens are involved in farming. Together these various criticisms of industrial farming have entered into domestic and global environmental discourses and are a motivating factor behind what I call religious agrarianism. Importantly, such critiques are encountered in the narratives and professed statements of religious agrarians from Koinonia and Hazon.

SUSTAINABLE AGRICULTURE/FARMING

Critiques of industrial agriculture presume that there is an alternative that is less destructive. This perceived better alternative is to be found in "sustainable" agriculture. Sustainable agriculture is popularly conflated with organic agriculture, while some supporters use terms such as *eco-agriculture*, *ecological agriculture*, *agroecology*, *regenerative farming*, or even the phrase "beyond organic."[36] Practitioners and supporters of such agriculture express a wide range of worldviews and engage in a diverse set of practices that are united by a belief that industrial agriculture is destructive socially and ecologically. They also share the belief that there is a better way to feed people and treat nature, and many believe that regional communities and economies are strengthened by supporting local, sustainable agriculture.

Advocates claim that sustainable agriculture is capable of feeding all people. This is because sustainable agriculture is made possible by adopting farming practices advocates claim are healthy for the soil, the crops, and animals, the larger environment in which these flora and fauna reside, and for human bodies and societies. Some of these farming practices include building soil nutrients and humus with compost; using GMO-free seeds and heirloom seeds when possible; the disavowed use of almost all petroleum-based and synthetic chemicals that are seen as being toxic and polluting; the humane treatment of animals raised for human consumption; the humane treatment of farm workers; and the development and maintenance of on-farm biodiversity. The last is made possible through planting polycultures and creating on-site habitat for life forms that are not directly part of the farm economy but are seen as beneficial.[37]

I will quickly summarize key attributes and goals of sustainable agriculture as articulated by leading exponents from the United States. The reader should note that for the practitioners and supporters of sustainable agriculture, these stand in direct contrast to the methods and goals of industrial agriculture.

Sustainable Agriculture: Land Health

This aspect of sustainable agriculture is articulated by the environmental philosopher Paul Thompson, whose 1995 book *The Spirit of the Soil: Agriculture and Environmental Ethics* is one of the earliest sustained analyses of the ethics of sustainable agriculture. He outlines the history of industrial agriculture, which he describes as holding to a "productionist" philosophical paradigm of sustained and increased yields and argues that this is the "sole norm for ethically evaluating agriculture" (1995: 47). For Thompson, this normative productionist ethic and paradigm contrasts with that of agricultural and agrarian stewardship, which is "both an ecological and a self-interested notion" (ibid.: 75–76).

Of direct import to the argument of this book is his point that

> Agriculture described in Judeo-Christian [*sic*] religious teachings is not typical of farming during the post-feudal era of concern here. [Rather,] stewardship is not something that farmers undertake altruistically, nor is it a religious duty that farmers perform at the expense of their personal, earthly betterment. . . . Stewardship is an integral component of agricultural land use. Stewardship duties do not oppose use, but are components of wise use. (ibid.: 74)

Here he eschews the romantic view of holism and "hands-off" environmentalism that has tended to dominate the majority of US environmental thinking over the past century, including in some religious environmentalist discourses about stewardship. He is arguing, rather, that farmers directly engage with and use farming landscapes for self-interested reasons that can nonetheless also be environmentally sustainable and that this practical tradition of stewardship has folkloric, ethical, and religious dimensions.

Thompson continues by categorizing agrarian stewardship as being "built on implicit or indigenous knowledge of ecology relating soil, water, plant, and animal life. These relations are understood as at least partially regenerative, and the good steward develops a working knowledge of how to enhance, rather than degrade, the regenerative capacity of soils and other farm ecosystems" (1995: 77). Notice his discussion of regenerative relations that are a hallmark of sustainable systems—sustainable agriculture attempts to regenerate and build the fecundity and health of agricultural soils and farm systems, holding that this leads to healthier and more nutritious foods, human bodies, and ecosystems (compared to productionist, industrial agriculture, which is seen as being degenerative).

Sustainable Worldview: Community Health and the Local

Thompson includes in his argument about stewardship a concern for the health of local landscapes, bodies, and communities. Another leading advocate of sustainable agriculture is Wes Jackson, founder of the Land Institute in Salinas, Kansas, where he and his staff have developed what he calls "natural systems agriculture."[38] Jackson is a trained botanist and plant geneticist whose 1985 book *New Roots for Agriculture* is an early emblematic call for sustainable agriculture that more fully expresses a concern about local health.

Like Thompson, Jackson begins with a critique of industrial farming, writing, "So destructive has the agricultural revolution been that, geologically speaking, it surely stands as the most significant and explosive event to appear on the face of the earth, changing the earth even faster than did the origin of life. . . . In this sense, then, till agriculture is a global disease, which in a few places has been well-managed, but overall has steadily eroded the land" (1985: 2).[39] He lambasts the productionist farming of the industrial paradigm and critiques the current focus on increasing yield as the sole determinant of successful farming. For Jackson this translates into the current US farm policy, which is to "simply discount the future" (ibid.: 21) based on a chemically intensive "chemotherapy on the land" (ibid.: 23).

This critique of modern industrial farming presents the context of Jackson's understanding of sustainable agriculture. Of import, he offers a vision of sustainable farming practices and ethics in contradistinction to the error-filled chemotherapy of modern industrial farming. He writes, "If we make this agriculture less human-dependent and more self-renewing, then the new agriculture, based more on the principles of nature, can afford us a greater opportunity to take without thought for the morrow and still be sustained. As we look to a new agriculture, we cannot nor should we separate our agriculture from our religion or from our ethics . . . as we move toward a sustainable agriculture, we will necessarily develop an ethic with sustainability at its core" (1985: 64). Two points stand out here. The first is that Jackson is arguing what Thompson argued ten years later—that our agriculture should be based on understandings of sustainable systems, including their abilities to be resilient and regenerate, and that our modern industrial agriculture is perceived to be antithetical to these abilities. The second point is that Jackson argues that sustainable agriculture requires a religious ethics of sustainability.[40]

Jackson's vision of sustainable agriculture is based on solar energy and privileges perennial fruits and seeds, with the latter coming mainly from grains. For him, such a sustainable system of low-input perennial fruits and seeds will be part of a system of farming that brings an "age of the recognition of limits" (1985: 122). Entering into this age will entail a shift in ethics and values toward our farms and systems of farming. This shift will tender a recognition that locality is important. It is important for land stewardship, a localized land ethic (the implications of which are investigated in Jackson's book *Becoming Native to This Place* [1996]), and concepts of robust health. This health will include a focus on local economies, farmland and farm soils, farm culture, human bodies, local waterways, and the bodies of plants and animals. Similar concerns permeate the writings of other advocates for sustainable farming, from Wendell Berry to Michael Pollan, and are found in literature on bioregionalism (which is further explored in chapter 4).

Sustainable Worldview: Spiritual Health

Although presented as separate subheadings, it is important to note that in the sustainable worldview, concerns about stewardship, locality, ecology, and individual health are interrelated. To explore the concern about individual spiritual health (which includes the health of those in religious communities), I provide a vision of sustainable agriculture seen in the work of the Protestant agronomist Gary Fick. His 2008 book *Food, Farming, and Faith* is a layperson's exegesis of the Hebrew Bible and New Testament, where an appeal to scripture is used to evidence the need for contemporary sustainable agricultural practices.

Fick begins by noting, "Sustainability requires change, and for people to change to more sustainable patterns of living, they must be knowledgeable and motivated. Faith or worldview is a great motivational force" (2008: xvi). Notice the privileging of faith and religion or ideas and worldview in the need to move toward sustainability (a move in the lineage of Lynn White Jr.), as well as the author's concern with sustainability in general.[41] Fick positions agriculture as a human activity that is culture bound and bound by nature and nature's limits.[42] This echoes Jackson and Berry, who challenge farmers to place their work within the context of localized limits. Therefore, Fick appeals to holism and underscores the crucial place of ethics and religion in analyses of food and agricultural issues.

Fick's reading of scripture leads him to claim that stewardship is the foundation of the biblical view of agriculture (2008: 17). Part of stewardship is local knowledge (ibid.: 54) and reverence for God's creation (ibid.: 55). Fick concludes in a vein that is similar to Jackson and Thompson: that systems thinking should inform our agricultural decisions and practices. Furthermore, religious and ethical beliefs, as well as farm labor issues and farm economics, are part of systems—and thus of a farm system—as much as tillage machinery and crop and animal breeds (ibid.: 127). Fick concludes his call for religiously informed sustainable agriculture with a list of what he calls "the essentials of agriculture" (ibid.: 179–80).[43]

Ellen Davis's *Scripture, Culture and Agriculture* (2009) is another publication that approaches agrarianism from a religious (in her case, theological) perspective. She claims that "In contrast to ourselves, [the ancient Israelites] belonged to a culture that recognized land care as the life-and-death matter it unquestionably is. Thus, they can provide a vantage point from which to view and develop a nuanced critique of our current cultural practices regarding land use and food production" (2009: 2). She maintains that the Bible has a theological land ethic (ibid.: 25–26) and this is especially found in Deuteronomy (ibid.: 27). Just as Emma Tomalin (2002, 2004) writes that a religious environmentalist view brings with it a reinterpretation of tradition, rather than a traditional interpretation, I argue that Fick and Davis take an environmental agrarian view toward Scripture. They both offer a reinterpretation of tradition, largely because environmental agrarian discourses and critiques of industrial agriculture are by definition only possible in a post–Green Revolution world.[44]

In the foregoing sections, we clearly see that leading, influential advocates of sustainable agriculture consistently underscore the moral and religious dimensions of alternatives to industrial agriculture. A common refrain and argument of this book is that this aspect of a sustainable worldview is gaining adherents throughout US religions, with Koinonia and Hazon providing leading examples of how religious ethics actively influence and shape sustainable farming practices.

Sustainable Worldview: Ecology/Natural Systems

Besides the focus on ethics and religion, advocates equally point to the need to understand how ecosystems work, point out that farms are managed ecosystems, and explicitly point toward the need for a holistic worldview within which sustainable farming is but one component. As with any activity that is managed and is part of a system, the values and ethics system managers bring into their decision making will affect the management and structure of the system. Thus, values and ethics shape systems, and thus affect, create, and shape the even larger systems (environmental, social, religious, political, economic, and others) within which these are embedded. Almost every critic of the system of industrial farming, and thus almost every proponent of sustainable farming, claims that the former is anthropocentric, mechanistic, reductionist, productionist, linear, polluting, and unsustainable. In contrast, they claim that sustainable farming is cyclical, regenerative, holistic, organicist (in worldview), adaptive, and harmonious.[45] The ethics and values of both systems are constantly seen to be at odds with one another and critics of the industrial model inform their criticisms by the holistic values and ethics they hold. Their worldview, and thus their idealized version of a sustainable planet, is directly threatened by the actions and values of the industrial system and worldview.[46]

Furthermore, for religious agrarians, a theistic understanding of creation provides motivation to criticize industrial agriculture and support the goals and ethics of sustainable agriculture. This is one of the key insights of this book: religious values, identities, and commitments provide a unique space for understanding and criticizing industrial farming systems. Although there is crossover with secular ethical and scientific critiques, as I am explaining and exploring, there is also something unique, different, and emergent in the religious agrarian critique. This unique emergence is based on a clear appeal to the divine/sacred and the effects industrial farming has on a holy creation.

The Return of "Place"

Farming practices can tell us more—not only about worldviews regarding ecology or the divine but also about place. Farming practices and philosophies reflect and reinforce the connection between religion and ecology and another connection—between ecology and place. Many people write about this, including Jackson, Berry, and Norman Wirzba. An astute philosophical exploration of place is found in Edward Casey's 1998 book *The Fate of Place*. He gives an intellectual history of the philosophy of "place" and how place has been conceived in the Western world since antiquity. He adds to this dialogue by noting that place as a philosophical concept has been given short shrift, as the West has fixated on space and time (Casey 1998: x).

Casey's argument about place analyzes ancient myths and religious narratives, Hellenistic and Neoplatonic thought, and early modern theories of place and space. He then concludes with modern thinkers who are on the vanguard of the reemergence of place over space and time in late modernity and postmodernity. He writes in this exploration that "Integral to the genius of early modern thinkers from Descartes to Leibniz is a disdain for the *genius loci*: indifference to the specialness of place, above all its inherent 'power'" (1998: 133–34).[47] This indifference to the parameters, boundaries, and uniqueness of place is one of the criticisms of industrial farming: the same farming technologies and methods are used to grow the same strain of seeds for the same international markets the world over, regardless of local social and environmental consequences.[48] Critics argue that this process is inherently destructive to land, farmers, farm economies, and human and nonhuman bodies. For example, Vandana Shiva argues that such a uniform privileging of the world (i.e., Casey's world of space and time) as interchangeable components of inert matter to be manipulated and traded as commodities leads to "monocultures of the mind" (1993). She argues that such monocultures in the mind generate monocultures of culture and landscapes that are damaging to women, ecosystems, and sustainable regimes of traditional subsistence agriculture.

Casey continues, offering in his conclusion that "Despite the seduction of endless space (and the allure of serial time), place is beginning to escape from its entombment in the cultural and philosophical underworld of the modern West" (1998: 339). My interpretation of Casey's argument is that this call for recognizing the inherent worth and value of place is one of the major goals driving sustainable agriculture. This concern is seen in the emergence of the Slow Food movement, of the Transition Town movement, of farmers' markets trumpeting local produce; of *terroir* (Trubek 2008), of the concept of "food miles," and finding and using local and heirloom seeds, as seen in the work of Seed Savers Exchange, for example. All of these submovements within the sustainable agriculture milieu advocate for the importance of a healthy, local place, where this "place" includes soils, what is grown on the soils, a reduction in distancing from the places where our food is grown, to even a specific, place-based taste that imbues locally grown food and value-added items. It includes movements toward place-appropriate, locally adapted varieties of flora and fauna, compared with the few species of globally traded "place-less" (and some argue tasteless) varieties of flora and fauna produced by industrial agriculture.[49]

It is also seen in the experience of sustainable farmers, as many argue that farming is an embodied act. In one sense, sustainable farming requires paying attention to local conditions and variables; in another, it requires the actual use of the body and its senses, limbs, and muscles to dig a trench, turn a compost pile, run a tractor, harvest a crop, prepare a booth at a farmers' market, shear sheep, and physically perform countless other tasks. This embodied action in nature is seen by many sustainable agriculture farmers (and, to be sure, some conventional/industrial farmers) to be salubrious, bringing with it wisdom, a sense of well-being, and a sense of meaning largely triggered by being directly involved with a cycle of life and death that is felt to be bigger than the individual farmer. If we remember that "for 350,000 generations humans have lived close to the land as hunter-gatherers [, then] a sense of belonging, place, and feeling embedded within the broader natural world" that is triggered by farming can be seen as a beneficial return to place (Mayer et al. 2009: 635).[50]

SUSTAINABLE VERSUS INDUSTRIAL AGRICULTURE

Given the foregoing aspects of and trends within a sustainable agriculture worldview, it is no surprise that many who hold this view have a strident critique of industrial agriculture. These include critiques that industrial agriculture relies on toxic chemicals and petroleum products; is inherently brittle due to overreliance on a few hybrid strains of food items; is undemocratic with a revolving door between corporate lawyers and employees and regulatory governmental agencies charged with overseeing agriculture; it disregards the welfare of animals and workers; and it produces food items devoid of flavor and nutrients (Minkoff-Zern and Getz 2011; Norberg-Hodge et al. 2001; Meghani and Kuzma 2010).

Meanwhile, advocates of mainstream, industrial agriculture claim that sustainable farming is unrealistic in its assessment of industrial farming and criticize sustainable agriculture for a variety of reasons. These include claims that sustainable farming is nostalgic and unreflectively opposed to progress, is unable to produce enough yield to feed more than 7.5 billion people (soon to be 9 billion); its crops are susceptible to rot, blight, pests, and drought because of its reliance on organic farming techniques; and last, that it is flawed in its overall claims of superiority because there is minimal to no scientific proof behind its varied claims about the benefits of sustainable agriculture, nor behind the claims about the ills of industrial farming.

The difference in worldviews between these two types of farming, including the views one side holds about the other, in turn influence values, emplaced bodies, and farm landscapes. They also can make for contentious political lobbying and uncomfortable neighbors, especially when views and thus practices occur within a bordered community. Just as sustainable farmers are involved in the production of place, so too are "conventional farmers . . . involved in the construction of places, only their places are quite different from those constructed by organic growers. And so we have two landscapes, two visions of community and two understandings of the world inscribed in the same geographical space. Therein lies the major site of friction between conventional and organic farmers" (Hetherington 2005: 30).

There are other criticisms against sustainable agriculture. Many are directed at USDA-certified organic products and their production and consumption. For example, much of the organic produce on US grocery store shelves are actually grown in Southern Hemisphere countries, especially during the North American winter, so these are export crops with a large ecological footprint.

Furthermore, by growing for export, local farmers in poorer countries do not grow enough food to meet their own caloric needs (Fair Trade products are one response to this criticism). Another common criticism of organic produce is that it is too expensive and elite, catering to a bourgeoisie foodie upper class. Such arguments point out that health food stores and other shops where organic produce is sold tend to be located in areas of affluence. As a result, there are "food deserts" in poorer areas where fresh produce (organic or industrial) is hard to come by. Issues of class and race tend to underlie this strain of argument.

Another criticism of sustainable agriculture is that not enough people are interested in becoming organic farmers. This means there will be a lack of willing farmers and people to train, and thus supply will never meet demand. Coupled with this is a charge that a reliance on smaller-scale polyculture acreage demands more workers and farmers and there are not enough of either to meet the needs of organic ideals. Last, many critics claim that organic has become a greener version of industrial farming. These critics point to the large monocultures of current organic farms while others maintain that organics have been taken over by corporate agribusiness interests.[51]

Another view criticizes both industrial and sustainable agriculture, especially if either forms are practiced at increasingly larger scales. This third view, with a broadly conceived understanding of agriculture as being dependent on settled communities of farmers who are engaged in domesticating a variety of plant species, is proposed by a minority of scholars, agronomists, and cultural critics. As already seen, Wes Jackson criticizes the reliance on till agriculture (as compared to no-till agriculture) and how this method of plowing soil leads to soil loss (a similar argument was originally made by Edward Faulkner in 1944). Another critique of human agriculture and the domestication of plants and animals and how this domestication reflects an urge to dominate nature can be seen in the work of Kirkpatrick Sale (2006). Meanwhile, Richard Manning (2006) and Paul Shepard (1998) criticize the human practice of agriculture, with each author sharing a similar argument about how human genes and societies evolved in small-scale gatherer-hunter enclaves so that with the onset of settled agriculture humans "fell" from an idealized, prelapsarian eco-past. Another critic is Weston Price, who traveled the world researching and reporting on indigenous diets and nutrition. He found that once food items from modern agriculture began to be consumed, never-before-seen tooth decay and other physical deformities entered into indigenous communities (Price 2006). Finally, one of the most popular and widely read criticisms of "totalitarian agriculture" comes from the neo-animist and communitarian Daniel Quinn (1992, 1996).

What these authors have in common is a critique of modern agriculture—its methods, technologies, underlying epistemologies, the effects on our view and relationship with nature, and the impact it has on the health of human bodies and communities. They also criticize the effect agriculture, especially modern agriculture, has on the health of soil, bodies of water, nonhuman animals, insects, microbial life, and the Earth's climate. Although most would be sympathetic to and would most likely promote sustainable agriculture over industrial agriculture (as Jackson does), they all criticize "agriculture" as a human technology and lifestyle. For these authors, humans made a deleterious and damaging mistake when we settled into permanent locations and began to actively replace native ecosystems with tilled agricultural fields planted with domesticated seeds and then populated with domesticated animals approximately 10,000 years ago.

By looking at industrial farming/agriculture, sustainable farming/agriculture, and criticisms of any form of settled agriculture, we are able to see how food and farming, and the values behind and wrapped up within them, are an important aspect of religion in the United States.

WHITHER AGRARIANISM?

Contemporary agrarianism is a response to modern industrial agriculture and governmental policies that support the industrial model of farming. Given this, contemporary agrarianism in the United States privileges sustainable farming as a hallmark of a responsible culture. The values that imbue a responsible culture of contemporary agrarianism include fidelity to farming lifeways, thrift, ecological literacy, concern for local economies, and the belief that sound, sustainable farming is a form of art. These hallmarks highlight the need for scholars of religion to connect the values expressed in agrarianism and those expressed in religiously based agricultural communities and/or religious communities that support agrarian values, as these synergies often overlap and can be mutually reinforcing.

For example, Christian theologian and leading agrarian writer Norman Wirzba writes that:

> Authentic agrarianism, which should not be confused with farming per se . . . represents the sustained attempt to live faithfully and responsibly in a world of limits and possibilities. As such it takes seriously what we know (and still need to learn) about the earth—the

> scientific ecological principles that govern all life forms—and what we know about each other—the social scientific and humanistic disciplines that enrich human self-understanding. (2003a: 4)

For members of the religious agrarian communities discussed in this book, these concepts of limits and responsibilities—coupled with deliberate and intentional living centered on their religious identities and their interpretations of their respective religion's teachings—help form a guiding vision and ideal that dictates their lived behaviors. These behaviors are reflected in and practiced by the members as they attempt to embody agrarian ideals of sustainability by supporting local food production and engaging in eco-friendly and environmentally responsible farming methods. Such behaviors—enforced, inspired, and reaffirmed by religious teachings, rituals, practices, and beliefs—are a phenomenon worthy of study.

To aid such study, I operationally categorize three phenomena (of many) that are compatible with and fit within larger contemporary agrarian ideals: organic (sustainable) agriculture, farmers' markets, and community-supported agriculture (CSA). In the United States, organic farming has grown exponentially over the past twenty years. This growth is financial, with sales of organic produce numbering in the billions of dollars, and applied, with the number of farmers seeking organic certification numbering in the thousands and the amount of acreage certified organic by the US government numbering in the millions.[52]

Various studies show that this increase in consumer demand for organic products is a result of health, labor, epicurean, and environmental reasons (Yiridoe et al. 2005). In response to this growing consumer market, some farmers have shifted to organic agriculture for economic and soil health reasons; some choose to farm organically because of ethical and spiritual reasons (Ableman 2005; Bolduc 2008). Besides this increase in organic food sales, there has also been an exponential increase in regional farmers' markets where fresh, local, and artisan food, animal, and craft products are sold. These products need not be organic, but most markets do have organic products present.[53]

The third recent phenomenon within a larger agrarian rubric is the bourgeoning CSA movement. CSA farms are those that offer "shares" to customers, where the customer enters into an advance contract with a farmer by paying up front for a share of the produce the farm will produce over the growing season. Such arrangements help farmers secure capital at the beginning of the season, when they most need it. They also signal that customers are willing to take risks with what the farmer is able to grow in exchange for receiving local produce on a weekly basis. CSAs can be organic or nonorganic and can include animal products; shares can be picked up on site or can be delivered (for example, some CSAs in New York state deliver to customers in New York City); and some have contracts where members must put in a certain amount of labor on site throughout the growing season. CSAs and farmers' markets represent one of the major goals and ideals of agrarianism: developing and supporting local food systems that recognize and operate within the limits of local "foodsheds." A further ideal and goal is to develop local food systems that are sustainable and contribute to ecosystem health; this is where organic/sustainable agriculture methods of farming explicitly enter into the agrarian worldview.

As noted earlier, during this same period of time, US religious communities, organizations, and institutions have demonstrated an increased environmental concern (Oelschlaeger 1994; Tucker 2003; Tucker and Grim 2014). This increase in environmental concern among some religious bodies does not necessarily correlate with the growth of organics as an industry and as farming practice or with the growth of CSAs and farmers' markets that support sustainable farming practices. However, some religious communities have chosen to express their burgeoning environmental concern by actively participating in or supporting organic and/or local farming practices. One of the key goals of this book is to explain and make sense of this phenomenon.

Despite the exponential increase in food knowledge at popular levels in US society (helped by the writings of Michael Pollan [2006, 2008], Joel Salatin [2007], and Wendell Berry [see later discussion] and films like *Food, Inc.*), the scholarly literature on the interaction between religious belief and agrarianism, especially sustainable agriculture, is sparse. One reason behind this dearth of literature is that social scientists tend to study either why individuals chose to farm organically or the growth of organics as a whole industry. Furthermore, studies about the motivations of organic farmers usually emphasize market forces or environmental concerns held by either farmers or consumers (Willer et al. 2008), rather than religious, ethical, and value-based motivations.

Within the field of religious studies, the few completed studies on sustainable agriculture have focused on individuals or communities (Gould 2005; McFarland Taylor 2007; Sanford 2012), but not a comparison of two or more religious communities, let alone similarities and differences between Judaism and Christianity.[54] There are theological treatises about organic farming (for example, Fick 2008; Kirschenmann 2010; and Bahnson 2013),[55] but these are written from a position of faith by members of a faith tradition and are not objective, comparative scholarly studies. A few scholars in the fields of religious studies and sociology are beginning to theorize about the influence contemporary environmental science has exerted on religious beliefs and practices (Taylor 2002: 49;

Proctor and Berry 2005); in general, this is still an underdeveloped field, and as a whole, the discipline of religious studies has tended to marginalize religion and agriculture issues.[56] Graham Harvey recognizes this when he writes, "Food and drink are not yet sufficiently part of our theorizing of religion" (2013: 30), such that my book serves to help continue filling this lacuna in scholarship.

Agrarianism holds a special place in the mythical and historical narrative of the United States, and its connection to modern religious environmentalism is the focus of this book. The agrarian mythos is directly built on the political philosophy and agrarian ideal espoused by Thomas Jefferson. The place agrarianism holds in the US domestic pantheon of social and political identity therefore equips the term with a certain amount of social capital now being used to build environmental, political, and religious networks. In fact, a key notion underlying the various agrarianisms of Western culture is "the idea that agriculture and those whose occupation involves agriculture are especially important and valuable elements of society" (Montmarquet 1989: viii–ix) such that "agrarianism may enjoy a significant future" (ibid.: 221).

Although the conclusion offered by Montmarquet's comprehensive study of agrarianism fails to predict the growing role religious environmentalist concerns (and environmental concerns in general) might come to play in a North American agrarianism, it nonetheless points toward agrarianism's continuing influence on the US body politic.[57] In his study, Montmarquet singles out Wendell Berry as the modern agrarian writer par excellence, justifying this claim based on Berry's prolific career writing about agrarian issues from a post–Bretton Woods, post–Rachel Carson, post–Earth Day perspective. Berry himself writes that

> We agrarians are involved in a hard, long, momentous contest, in which we are so far, and by a considerable margin, the losers. What we have undertaken to defend is the complex accomplishment of knowledge, cultural memory, skill, self-mastery, good sense, and fundamental decency—the high and indispensable art—for which we probably can find no better name than "good farming." I mean farming as defined by agrarianism as opposed to farming as defined [pace Thompson, Jackson, and Fick] by industrialism: farming as the proper use and care of an immeasurable gift. (2003: 24)

Berry's position that agrarianism is art rests on his conviction that an art-based agrarian culture is a derivative product of the land. For him, the art of agri-culture is based on a healthy relationship with the land on which all food is grown. Good farming mixes culture, knowledge, restraint, hard work, and neighborliness and places this mixture onto local fields and into local forests.

The agrarian—and for this book especially the ecological agrarian—focus on the local is echoed by Thomas Hilde, who writes that, "One thing all versions of agrarianism share . . . is a conception of the land as much more than a material resource" (2009: 20). This insight is seen in agrarian writing from the early 1900s, for example, in the work of Troy Cauley, who states, "Agrarianism may be roughly and tentatively defined as an economic and social system under which the chief method of making a living is that of tilling the soil, with a consequent rather wide dispersion of population and a relative meagerness of commercial intercourse. It is, probably, *the antithesis of Industrial Capitalism*" (1935: 3; emphasis added).

Given these definitions, it is easy to see how agrarianism fits within the lineage of and appeals by those US citizens who are supportive of environmental issues, especially issues related to sustainable agriculture. However, this appeal to environmentally concerned citizens and devotees of sustainable agriculture was not originally evident in the US agrarian tradition. For example, W. B. Bizzell comments that, "*agrarianism signifies an organized effort on the part of the farm population, or a socially conscious group of farmers, to secure a redistribution of land or the establishment of law of conditions more favorable to the use and occupation of land.* An agrarian revolution is concerted action on the part of farmers to bring about economic or social changes that promise to improve farm life conditions" (1926: 2; emphasis in original). Although we see here a history of social and economic critique (but note, and this is key, minus a sustained environmental critique) and concern for the livelihood of farmers, this history has been in reference to particular events, epochs, and regions within a trajectory of US history. This includes epochs when the majority of those in the United States (and thus voters) were active farmers.[58]

Even though Bizzell comments on policies of the early republic that shaped later agrarianism—notably the transfer of vast tillable land from public to private domain, the rapid increase in population, the use of increased machinery and farm technology, the import of new livestock breeds, the development of transportation and facilities, the extension of farm markets into growing cities, and the rise in scientific knowledge about agriculture (1926: 10)—these trends reflect US politics of the late 1800s and early 1900s. While contemporary agrarians criticize these same processes, today's agrarians must also contend with post–Green Revolution technologies, greater industrialization and mechanization of farming, a more apathetic public that has significantly reduced its number of working farmers, and international agricultural markets that are controlled by agribusiness corporations.[59]

To dig deeper into why agrarianism has a continued and even growing foothold in certain sectors of today's United States, we need to follow Bizzell's and Cauley's expositions about the roles industrialism and capitalism played in shaping modern farming. Indeed, this exploration joins streams with the varied criticisms that modern champions of sustainable agriculture have of capital-intensive, industrial farming. For example, a thoroughgoing critique of the machinations of industrialism is found in the rise of English radical agrarianism of the late 1700s and early 1800s (Chase 1988), and this critique helped shape some of the agrarian ideals and narratives in the late 1800/early 1900s United States. This is especially evident when Chase writes, "Agrarianism was centrally concerned with skill, security, independence, and status . . . So, far from being some retarded throwback of preindustrial, 'traditional' society, agrarianism can—and indeed must—be located precisely in the experience of labour during industrialization" (ibid.: 8).

Similar themes can be deduced from today's agrarianism, especially given that modern environmental agrarians are Berry's lamented "losers" in the ongoing process of industrialization and global enclosures of agricultural lands and markets. This means that although environmental (see below) and religious agrarianisms developed within the politics of the United States, it did and still does argue against this now pan-global history of agricultural industrialization and loss of farm labor and culture.[60]

Further international links are embedded in today's agrarianisms. For example, the US permanent agriculture movement of the 1930s and 1940s was influenced by the writings of Sir Albert Howard, whose famous treatise helped spawn the "organic" agricultural movement (and played a major role in the development of Berry's agrarian ideals). Howard himself was heavily influenced by observing the various regional farming practices of India. Of these practices, one of the most important for him was witnessing and measuring the effects of the copious amounts of compost Indian farmers added to their soils.

Scholar Kim Smith provides us with an important link between these past agrarians and the dominant agrarianism of today, or what she calls "environmental agrarianism" (2003). This most recent strand of the agrarian tradition provides the major taproot on which religious agrarianism is grafted, so it is important to understand the development and worldview of environmental agrarianism.

Smith devotes much of her monograph to highlighting the importance of Wendell Berry, who in her interpretation is the iconic founder and exemplar of modern environmental agrarianism. As she states, "Berry is a central figure in the greening of American agrarianism. Since the 1960s, he has been a leading expositor of a set of ideas designed to forge a politically effective union of small farmers and environmentalists" (2003: 1). His agrarian thought is consistent with and builds on earlier agrarian concerns about the effect industrial capitalism has on farm communities and farmlands, and thus on society as a whole. However, she notes that his agrarian vision depends on a rejection of traditional agrarian stalwart concerns, such as "individual independence, the sanctity of property rights, and the meaning of economic freedom" (ibid.: 2–3) and instead offers "a provocative blueprint for an alternative, ecologically sensitive agrarian society based on the value of stewardship" (ibid.: 2–3). Thus, according to Smith, Berry's writings become the starting point for a nuanced, introspective, culturally powerful, and sustained environmental agrarian criticism of modern industrial capitalism, as well as the political, economic, and agricultural systems it has spawned.

Many if not all of the other books on agrarianism that I have read while writing this book all point to Berry's significance in creating this environmental agrarian tradition. His insights largely permeate and influence religious agrarianism. Both religious and environmental agrarianism recognize that "Good land use is likely to occur . . . only when it stands as a shared aspiration. Society must embrace the goal and work toward it" (Freyfogle 2007: 2). Because religion is part of society and values represent aspirations, it becomes evident that religious values (including religious environmentalist values) are now mixing with sustainable agriculture and environmental agrarian values. The mixing of these values is changing the US food culture. As Freyfogle states, "A culture writes its name on land for all to see. . . . The land reflects not just what people have done but who they are, what they understand, what they value, and what they dream" (ibid.: 1). Given this, it is a mistaken assumption to disregard the power and influence that religion has on shaping US culture, and thus the landscapes and the values contained therein.

Such importance suggests that an environmental agrarianism for the twenty-first century contains the following values, ideals, claims, and attributes:

- It is both art and practice.
- It is essential to a healthy society, culture, and environment.
- It embraces technology, when appropriate, and places the use of technology within the limits and standards of the land and the communities (human and nonhuman) that are dependent on the land for their survival—therefore, the use of science and technology is a

derivative of and is shaped by the needs and constraints of local land bases and local ecosystems.

- It views land, especially the soil, as more than a mere inert resource for an extractive economy; rather, land and soil are seen as the basis of the total human economy.
- It takes stewardship of the land seriously, to the point where stewardship is one of the key driving values and ideals of a modern, ecologically informed agrarianism.
- The practice of an environmental agrarianism is a learned skill.
- It is a comprehensive worldview.
- It criticizes the short-sightedness of the extractive, industrial economy, including especially productionist agriculture.
- It can and often does have a spiritual and/or religious component.
- It values democracy and advocates people taking control of their food and thus of agriculture and culture.
- It argues for propriety and familiarity with local lands, people, and customs.
- It critiques dominant land use laws and patterns (whether local, national, or international) and the free-market ideologies that support these.
- It is built upon a long-standing historic tradition and lineage (especially political) within Western culture.
- It develops and promotes systems thinking, including the role humans, farm communities, agricultural markets, and human customs play within various interlinked ecological, cultural, and economic systems.
- It is holistic.
- It privileges the vocation of farmers-as-stewards (versus seeing "good farming" as a backward, ignorant, culturally taboo [especially for educated Caucasians] profession).
- It is open to building networks with others (nongovernmental organizations, lobbyists) at national and international levels for advocating and promoting agrarian and larger environmental concerns and ideals.

With this overview in place, coupled with my earlier exploration of religious and environmental values, networks, and lived religion, I am ready to explicitly develop what I mean by "religious agrarianism." In chapters 2 and 3 readers will be introduced to the two key religious agrarian communities that embody many of these attributes of environmental agrarianism, adding to them religious concern for the local, health, and justice.

RELIGIOUS AGRARIANISM

Wes Jackson writes, "In the earliest writings we find that the prophet and scholar alike have lamented the loss of soils and have warned people of the consequences of their wasteful ways. It seems that we have forever talked about land stewardship and the need for a land ethic, and all the while soil destruction continues, in many places at an accelerated pace. Is it possible that we simply lack enough stretch in our ethical potential to evolve a set of values capable of promoting a sustainable agriculture?" (1985: 13). In many ways, this quote is central to the religious concerns this book investigates. Religious agrarianism is a developing set of practices that brings explicit religious environmentalist concerns, values, and ethics into the perennial problems of soil destruction. This destruction has become amplified since the onset of the Green Revolution, and the environmental agrarianism of Berry et al. represents that tradition's most up-to-date attempt to deal with the issue. As we have seen, largely since the publication of Lynn White Jr.'s 1967 article have people acting from within various religions started to take environmental and ecological issues seriously, adding these issues to other longer-standing concerns.

Such environmental concern has been slow to develop, in no small part because White specifically targeted Western Christianity as the culprit of said crisis. The earliest responses from apologists (whether from inside a tradition known as ecotheology or from scholars of religion) to these charges looked into the sacred texts of their various traditions (or traditions they studied), searching diligently for green passages as evidence of environmental responsibility to counter White's argument. Such a defensive ecological hermeneutic has come to be called the "mining" approach to religion and nature issues. An ideal type of such an approach is HarperOne's New Revised Standard Version of the Bible, *The Green Bible* (2008). This version is printed on Forest Stewardship Council–certified paper, has St. Francis of Assisi's "Canticle of the Creatures" as the first prayer on the opening page (possibly as a nod to White, who stated that St. Francis was the Christian eco-saint par excellence), and tellingly includes poetry from Wendell Berry.[61] Lest one think I am singling out Christianity by sharing this example, such mining has been and is still being done in most subtraditions of current religions, from Judaism to Islam, from Hinduism to Buddhism, and beyond.

The growth of religious environmentalism has also been inspired by seminal events like Earth Day, which began in 1970; the publication of *The Limits to Growth*; the Rio Earth Summit; and the growing ecumenism surrounding religion and ecology and evangelical creation care that have contributed to an accelerated growth of religious environmentalist concerns.[62] These concerns, along with a more sophisticated understanding of food politics and sustainable agriculture issues, are leading some religious environmentalists to embrace environmental agrarian concerns and values. As Gary Fick shares, "agricultural sustainability is inherently holistic. It includes all of the essentials of agriculture. It includes the past, the present, and the future. It includes all of the social, cultural, and ethical linkages in the food system as well. Thus religion is a natural component of holistic, sustainable thinking, and academics like me have erred when we have left it out of the picture" (2008: 6). Not only have academics left religion out of the sustainable agriculture variable, many of those active in religious networks have left sustainable agriculture out of the religion variable. Religious agrarianism is a developing corrective being embodied and put into practice by those who reside within religious traditions. The movement also allows a growing number of those living in the United States to equally embody and practice environmental agrarian concerns.

Religious agrarianism places the practice of sustainable farming into a larger, theocentric worldview that sees Creation as being holy and worthy of reverent care.[63] Religious obligations and duties follow from such a perspective, so that sustainable agricultural practices become a way that religious agrarians can live out their religious concerns. As Paul Thompson notes,

> The traditional agrarian view of stewardship can be summarized as a religious duty to protect and foster the beauty and integrity of God's creation. The primary flaw can be summarized, as well. Traditional agrarian stewardship is conceived as a duty ethically subservient to production; hence when stewardship would entail constraints on production, duties to nature seldom prevail over the productionist ethic. As such, if stewardship is to serve as a component in an environmental ethic for agriculture, it must be broadened and reshaped. (1995: 72)

Religious agrarianism provides this exact paradigm of a broadened and reshaped stewardship ethic: because the religious agrarians I encountered identify the land with a theistic Creator and because their values, lives, and ideals are shaped in such large part by their religious beliefs and networks, these agrarians represent a new addition to contemporary sustainable agriculture and environmental agrarian concerns.

Religious agrarians add to and expand the political, social, and cultural components of contemporary agrarianism. Kim Smith points out that there is a democratic agrarianism that has informed US politics and has focused on the values found within family farms. Religious agrarians do not necessarily make this exact claim, but some argue that ecologically informed religious communities that support sustainable agriculture/agrarian/environmental justice issues are such a repository of democratic and social ideals that are needed to combat industrial capitalism. Religious agrarians share many of the same critiques of industrial farming and capitalism that environmental agrarians develop—the difference is that these critiques are developed out of specific religious and ethical traditions. This means that religious agrarians bring religiously based ethics and concerns for justice and equality to their sustainable agriculture values (see especially chapter 6).

Overall, there are two dominant strains feeding into religious agrarianism. One strain pulls on the history of agrarianism proper, especially modern environmental agrarianism as exemplified in Berry's writings and the values, goals, and ideals found in the list I created. The other strain is an amalgamation of the unique religious traditions, sacred texts, practices, rituals, teachings, and modern institutions within which religious agrarians reside. Both strains contain critiques of modern industrial capitalism, and both see the land and soil as having intrinsic value. In the case of religious agrarians, this value is theistic and thus carries religious duties that are entirely consistent with the goals of environmental agrarianism.

Given the foregoing, it is salutary for me to address a criticism of my research and historical framing of agrarianism. This criticism also extends to religiously influenced back-to-the-land movements that include agriculture broadly. Some of this history has already been covered, and more will be covered in future chapters. Here I would like to take a moment to respond to a point raised during an early version of this book. A generous reader claimed that they were not entirely convinced that there is something new and unique about today's alternative food movement, let alone the interaction between religion and agrarianism. They pointed out that back-to-the-land movements have been a continual presence in North America, dating back to the 1800s, and can be seen in Mormonism, Christianity, and alternative religious/commune movements spread throughout the landscape of the United States.

My research shared in this chapter and the book strongly suggests that there are clear differences between religious agrarianism and these antecedents. For one thing, many in religious agrarianism are motivated by concern for climate change, species diversity, and farm worker health and recognize we are at peak oil and what I call "peak industrial agriculture." These are all new concerns, borne of a more robust environmental science that was not available prior to the 1970s. In essence, these contemporary concerns reflect a much larger human impact on the planet.[64] Also, earlier religiously inspired farming and back-to-the-land movements tended to be in the minority and had a strong countercultural ideological component. Compared to this, religious agrarianism today is more mainstream, as sustainable food concerns are also more mainstream than at any point in US history. This is largely due to Berry's writing, who begins the ecological agrarian tradition with his critiques of industrial agriculture. This strain of agrarianism is what religious agrarianism is in conversation with the most, and it did not exist prior to Berry's corpus. Industrial agriculture is also recent, coming out of the Green Revolution and World War II. Although there is indeed some level of continuity between and among traditions, as is to be expected in any religio-social movement, there are also unique and new political, economic, social, and environmental conditions that religious agrarianism is a response to/that shape religious agrarianism.

I approached Wendell Berry with this same criticism, and in a letter postmarked November 28, 2011, he wrote to me,

> I think your [reader] is right in calling attention to the historical precedents of the present day "alternative food movement." But this movement is fairly new insofar as it is ecologically based and is a response specifically to the abuses, and I would say the failure, of industrial agriculture. "Unique," I think, may be an unfortunate word. Whether or not the alternative food movement is unique is really of not much importance. What is important is to describe it accurately, and accurate description would require you to specify the ways in which it is new or, fairly new.

I am comfortable, and indeed the evidence seems overwhelming, with labeling religious agrarianism as something new and unique. From my viewpoint as a scholar, this comfort derives for the reasons listed in this chapter and explored throughout the rest of this book, where I attempt to describe accurately why this is so.

I continue this work in the next two chapters, where I introduce the two case study communities of this book and the networks of which they are part. Chapter 2 gives an in-depth exploration of Koinonia, followed by a chapter that delivers an equal treatment to Hazon. In subsequent chapters I compare and contrast their religious agrarian themes and concerns about locality, health, and justice. I focus on specific narratives and emergent themes from those I met in the field that highlight their religious reasons for becoming involved in religious agrarian lifestyles. Such an ethnographic focus underscores the steadily emerging, religion-in-everyday-practice-and-concern, bottom-up ways that religious participants in the United States put their values into practice in regard to sustainable agriculture issues.

CHAPTER TWO

KOINONIA AND CHRISTIAN RELIGIOUS AGRARIANISM

LAY OF THE LAND

The next two chapters introduce the people I encountered who are creating religious agrarianism. They do so via religion-in-action, embodied sustainable agriculture practices occurring in specific places, and generating religious agrarian values via lived networks. Each chapter provides key historical background and larger institutional contexts. I begin with Christian religious agrarians in the United States, with an explicit focus on Koinonia and its founder, Clarence Jordan. Without an understanding of this founding figure, as well as the community's institutional and communal history, it is difficult to place Koinonia and its permaculture practices within the emergent phenomenon of religious agrarianism. I explore what permaculture is so that in succeeding chapters we can see how it enables Koinonia to engage religious agrarian topics of locality, health, and justice. A brief history of sustainable farming within the larger tradition of Christianity is also offered, because this helps contextualize Koinonia's place within a lineage of human–agricultural interactions that have been shaped, broadly speaking, by the Christian worldview.

A DEMONSTRATION PLOT FOR GOD

The drive to Americus takes the traveler deep into southwestern Georgia and provides a snapshot into rural life in the upland South. There are Mennonite communities, African American Baptist churches, Methodist churches, migrant communities of Latino farm laborers and their respective churches and *tiendas*, trailer parks with more than just a hint of poverty, and miles of agricultural monocultures. The latter range from industrial chicken and hog farms to the ubiquitous regional pecan groves to acres of soybean, cotton, and corn.

The town of Americus was founded in 1832, and today it has become a pilgrimage destination of sorts, for it is where Habitat for Humanity maintains its corporate offices.[1] The two colleges in Americus, Georgia Southwestern State and South Georgia Technical College, attract a number of young adults to the area. Coupled with the staff, volunteers, and visitors who pass through to visit Habitat or stop on their way to Plains (home of former President Jimmy Carter), these students help create a vibrant cultural life in this otherwise rural, historically agricultural community. The closest interstate is over an hour away, with the closest city, Macon, being a two-hour drive and Atlanta a three-hour drive. Both are northeast of Americus, so to outside observers the town is clearly more rural and poor than urban and diversified.

This rural, agricultural identity and Jim Crow community dynamics attracted Clarence and Florence Jordan to move to a swath of run-down farmland outside of Americus to put their theological vision into practice and begin Koinonia in 1942. At that time, Americus was even more rural and more dependent on agriculture for its identity than the town of today with its colleges and visitors to Habitat. In 1942 the Jordans, along with friends Martin and Mabel England, "went to Sumter County and there, alongside Route 49 about eight miles southwest of Americus, were 440 ordinary-looking acres of soil, slightly eroded and virtually treeless. Through some power of hidden persuasiveness and sense of rightness that passes between man and earth, Clarence said: 'This is it.' " (Lee 1971: 33).

CLARENCE JORDAN

Jordan came of age, physically and spiritually, in the pre–civil rights South. Born into a Baptist family on July 29, 1912, he began at a young age to question the teachings and hymns of his Baptist upbringing that preached equality before God. His experience as a youth of the reality of Georgia at the time was that there was blatant racism inside the church doors on Sunday and outside the church every other day of the week. Jordan noticed the inequalities that resulted from the landed aristocracy, or "good 'ol boy network," that kept both low-income white and black farmers poor and indentured.[2] For this reason, he decided to become a farmer "and to try to lift the awful burden from the poor man's back by showing him how to get a lot from a little land. He would seek to work in partnership with the poor farmers. A volatile mixture of guilt and a rage for justice was stoking up a lifetime of pressure" (Lee 1971: 10).

This conviction led Jordan to matriculate at the Georgia State College of Agriculture at the University of Georgia. He also became a member of the Baptist Student Union and began preaching in local churches. While at agricultural school he joined the ROTC and was scheduled to enter the US Cavalry. However, his background with the Bible, especially the Sermon on the Mount found in the Gospel of Matthew, led him to become a pacifist and walk away from armed service. Instead, Jordan entered the ministry and became a New Testament scholar at the Southern Baptist Theological Seminary. In essence, he became a soldier for Christ fighting for racial and economic equality.

By 1942, Jordan had enough firsthand witness of "the nervous eyes of the leather-skinned rednecks who saw it as their community responsibility to keep the Negro iced with fear" (Lee 1971: 3). He was tired of witnessing "the day-to-day routines of white Christians who would not even acknowledge the Negro as fully human" (ibid.: 3) and who saw "negroes" solely as farm labor, and cheap farm labor at that.

While Jordan's Southern white counterparts were heading to Europe to fight World War II, he realized he had a fight of his own to take up in the rural South:

> He was determined to fight fascist like oppression in Georgia with something that southerners were almost as familiar with as they were their guns: Christianity. The difference was, Jordan did not talk about the "Gee-zus" of the feverish radio evangelist, or the soul-saving Jesus of the crude highway sign, or even the slick Jesus of the sanctuary. He talked about Jesus the man, as if the guy actually worked and sweated, experienced love and hurt, and had about him all the shrugs and shuffles of a down-to-earth human being. (Lee 1971: 4)[3]

Motivated by his interpretation of the biblical Jesus and "preaching that the ideas of the New Testament either had to be incarnated or rejected" (Lee 1971: 5), Jordan "established an experiment in Christian communal living on a farm in Sumter County, right in the heart of southwestern Georgia, and declared brotherhood, nonviolence, and economic sharing to be its fundamental guidelines" (ibid.: 5).[4]

As historian Charles Marsh explains, "Jordan believed that the only way authentic change could transpire in southern race relations was through 'incarnational evangelism,' and that meant making Christian truth concrete in community and in shared life with the excluded and the oppressed" (2005: 56). He saw "life in the body of Christ from the perspective of Sumter County, Georgia, [and this body] involved three interconnected passions: the practice of nonviolence as the moral disposition of the Gospel; the preservation, cultivation, and protection of the soil, 'God's holy earth'; and the proclamation and provision of hope to 'those who suffer and are oppressed'" (ibid.: 69). Given his cultural background in poor rural farming communities, his professional training in scientific agriculture, his theological training and beliefs in incarnating Christ's early egalitarian communities of the faithful and support from his wife, Jordan decided to take action and founded Koinonia to bring change to the South and to the United States at large. By building a "demonstration plot" of advanced farming practices, interracial living, and Christ-centered community life, he hoped to inspire other Americans, Christian or otherwise, to overcome the abject poverty and racism of the rural South.

To this end, Jordan began farming at Koinonia by hitching himself to a plow, as the community did not have enough capital in the first years to purchase a steer or ox. He launched a soil conservation program of terracing, used ground-up peanuts as fertilizer, established a "cow library" and egg and seed cooperatives, shared farming tips with neighbors, and held seminars on farm-related topics for both races.[5] From its beginnings Koinonia has had a theology of the soil and has attempted to put into practice progressive, sustainable farming techniques, all undertaken within the context of rural Southern poverty and embedded racism.[6] As with religious communities everywhere, Koinonia's history of practice has not always lived up to the ideals of its visionary founder.

KOINONIA IN THE 1960S

Koinonia's history as an intentional religious community is as rich and textured as the red clay soil found in its fields. As with any Christ-centered intentional religious community, there have been constant issues with attracting and maintaining committed members and with remaining true to incarnating the teachings of Jesus. Koinonia has continually struggled with developing and maintaining on-site cottage industries for economic solvency and successful farming practices. The community has also tragically had to deal with violent responses from the Americus community based on Jordan's interracial vision. Except for the violent responses, Koinonia still grapples with the other issues on a daily level.

Koinonia dropped to just four members in 1967, with the Jordans and the Wittkampers being solely responsible for a 1,100-acre farm and pecan business. This led to deep soul searching by Jordan, and his friendship with Millard Fuller (who moved to the community with his wife during these lean years) helped pull him through. The two of them focused on combating the US ideal of competition and getting ahead (in effect prefacing Bellah's critiques of rampant individualism), eventually beginning a program of providing quality housing to low-income African Americans who lived in the area.

This war against "the matter of land tenure" (Lee 1971: 208) led to the creation of Habitat for Humanity and highlights Jordan's commitment to an agrarian ideal. For example, a 1935 tract on agrarianism states, "A true agrarian economy is characterized by a general diffusion of property ownership. Each family should own outright a sufficient quantity of land and equipment to afford it a decent and satisfying living with the application of reasonable effort and foresight" (Cauley 1935: 108). Jordan shared this agrarian vision and wedded it to a theological interpretation of the book of Acts. Jordan even proclaimed, "We need to return to the Old Testament idea that the earth is the Lord's; it's still His property—no man can give you a clear title. If you trace the title back far enough, somebody stole it" (Lee 1971: 209).[7] It can be argued that Jordan and Fuller's Fund for Humanity, which they began in 1969 (which Fuller renamed Habitat for Humanity when he left Koinonia), prefaces today's micro-credit lending movement. Rather than giving out high-interest loans so local residents and poor sharecroppers could not afford a house, Jordan and Fuller allowed poor African American neighbors to move into homes built on community property under interest-free loans.

Jordan lived long enough to see the civil rights era come to fruition, and he shared thousands of meals with interracial groups on the farm he helped create. He died in 1969 while working in his one-room writing shack that still stands among the pecan groves he helped plant. These pecans provide Koinonia with the majority of the community's income, fueling a mail-order business that led to the construction of on-site processing, baking, and packaging plants that still function today.

KOINONIA FROM THE 1970S THROUGH THE 1980S

Before Jordan's death in 1969, he and Fuller had changed the name from Koinonia Farm to Koinonia Partners and began to emphasize communication, teaching, and application as the core focus of the community (today the community refers to itself as Koinonia Farm).[8] Koinonia also became incorporated as a nonprofit during this era and developed a board of directors and bylaws. The community began to allot a living allowance to all members with income largely coming from produce and pecans.

Given the focus on communication and teaching, leadership at Koinonia in this era began to focus on clearer communication among members, between the community and Americus, and from the community to the world at large via a pecan product catalog. They taught about farming, community life, and Jordan's *Cotton Patch Gospels*, and began applying these teachings to a life lived in fellowship. This clarity of focus led to "an energized community [of] new visions, ventures, and variations" in the 1970s (Castle 2007). A "fabric of partnership" (ibid.) developed based on Jordan's vision of living in a shared, Christ-centered community. This partnership was seen in housing, agriculture, and business ventures.

Regarding agriculture, over its history Koinonia has grown row crops (mainly soybean and peanuts), pecans (with shells being spread back over the soil to make compost), muscadine grapes (for wine and jams), and a variety of other crops. The farm flourished with the onset of Jesus and back-to-the-land movements in the 1970s, providing a stark contrast to the four members of 1967. During this time many structures were built on the property, and many committed pacifists, hippies, and Christians came to live and raise families and try to incarnate Christ's love by living in a farming-based community. The farm grew corn, peanuts, squash, and okra and raised pigs for farmers' markets and domestic use. Sewing and tie-dye enterprises were also created.

Despite this energy, the seeds for community conflict were already sown. Although embodying a partnership vision of blacks working and eating with whites, there was a trend of whites coming to live at the community but very few blacks doing the same. Many of the African Americans involved were nearby residents or those who had homes from the Fund for Humanity and who walked to work. Although all received paychecks, not all were equally committed to partnering in business and housing. As Castle writes, "This situation would plague Koinonia through the '80s and come to a head in the early '90s as part of an 'unfinished reformation'" (2007) when issues of race, income, and embezzlement affected the leadership and structure of the community.

Will Wittkamper, who had been involved with the community in some capacity since 1953, led the farming during this era, and he instituted organic farming practices. Some of these practices were organic by default, in that the community did not have enough money to buy large amounts of chemicals.

The community was also inspired by the back-to-the land ethos of the time and benefited from abundant farm labor.

Community life during this era faithfully held to Jordan's vision: each day began with prayer and a worship period at 8 am and a brief devotional was given at the shared noon meal. However, "Work is an excuse to be together as much as to create a product. The spiritual experience in relationships seems very important. Whether work at Koinonia is primarily to be a business or an activity for spiritual growth has always been in tension" (Castle 2007). This tension and the tension between those working for a paycheck (who tended to be local blacks) and those working in partnership (who tended to be white) began to surface in the 1980s and boiled over in the 1990s.

By 1980 Koinonia had twenty-four resident partners, eleven children living in the community, and many volunteers. Castle depicts three threads that bound the community during this era: an inward focus and search for deeper spiritual life; the maturation of outward ministries, especially the growing success of Jordan's Cotton Patch Gospels; and at the end of the 1980s the surfacing of an "unfinished racial reformation" (Castle 2007). The community suffered another loss during this era when Florence Jordan, co-founder of Koinonia, died in 1987.

The community also became very active in antiwar demonstrations and in supporting the rights of Latin Americans suffering under Reagan-era policies. Active campaigning against the training of soldiers in Ft. Benning, Georgia, at the School of Americas began in earnest and continues to be a focus of Koinonia members to this day. The community also underwent a sustained period of institutionalization with the creation of handbooks, guidebooks, living allowances, and requested vacation time. Many original buildings were torn down and new ones were built. And the onset of a "dark night" (Castle 2007) took hold as the 1980s turned into the 1990s.

KOINONIA IN THE 1990S

David Castle shares that in the 1990s

> Koinonia gave up the communal, common purse for a non-profit organizational model. Even though the economic pattern changed, many still referred to Koinonia as an open intentional community. It was open in the sense that visitors could come and go and Koinonia did not maintain a doctrinaire statement of belief that people had to accept in order to be a part of the community. This would keep Koinonia from being identified with the more staid, inflexible, cultic

> organizations. Jordan's belief in a more experimental, free-flowing, creative milieu was maintained. (2007)

This free-flowing milieu allowed for underlying racial tensions to boil up and for more vocal black workers to argue that if Koinonia wanted to walk its talk, then an African American should become the executive director of the incorporated businesses and community. Such a position was created beginning in 1994. While the main ministries of the community were building houses, advocating for prison reform, and working with children and youth, there was internal dispute over how the community should be structured and who should be in charge.

One issue of contention was that local blacks who worked for Koinonia as an incorporated business wanted the life that white community members were voluntarily giving up when they became community residents. A common phrase arose among black workers: "We've learned how to make everything out of nothing" (Castle 2007). These workers felt that Koinonia should pay better salaries and that an African American should be in charge of directing the community. To deal with this tension, the board of directors appointed a six-member committee in September 1992 that was charged with creating structural changes that would appease the concerns of all members. The board decided that everyone would become a partner of the community and get paid, but there would also be "resident partners" who lived full-time on community property. It also came to pass that a local African American would become executive director of Koinonia. By 1999 Koinonia was in a sizable and dangerous financial hole of $744,000. It turns out that the executive director had been embezzling funds, and this had placed the community in a state of financial and emotional crisis.

Koinonia members called 1999 the Year of Recovery. Outside consultants were brought in to help the community overcome its financial debt. With their input, community members decided that 800 acres would be sold, so today Koinonia has 573 acres of land of which 85 are tillable and another 90 are in pecan groves. New leadership was found as the old director had been fired, and financial record keeping became more transparent. Last, Koinonia remained active in its mail-order pecan business and still advertised itself as a place, a spirit, and a ministry.

KOINONIA IN THE 2000S THROUGH TODAY

The community's fortunes changed after 1999. The terrorist attacks of September 11, 2001 affected the community in particular ways, especially the militarized response of the US government and its invasion of Iraq. This military campaign touched a nerve in Koinonia, helping mobilize the community after the tumultuous 1990s. Through its history, Koinonia has been committed to the ideals of Christ-inspired pacifism; this was evident in the historical ministry on race and in protests and campaigns against the School of Americas. After George W. Bush ordered the post-9/11 invasion of Iraq, members of Koinonia once again became active in local and national peace campaigns and vigils.[9]

Another important event occurred in 2002, when community bylaws were changed. Eight people began to sit on the board of directors, and David Castle was made chair of the board. Also in 2002, John Hall became coordinator of operations. His background as a dairy farmer in Vermont helped the community focus on farm maintenance and upkeep. Another important event happened in 2002: Koinonia once again became financially solvent. Finally, a new partner covenant was created with eight covenant ideas that were drawn up to guide the vision of the community. These included a covenant that the community would "be a voice for peace, nonviolence, and social justice" in the larger community; that community members would "work toward making home, church, and community a 'demonstration plot for the Kingdom of God' "; that the community would "strive toward living more simply, recognizing the terrible inequities of wealth, healthy, property and opportunity" (this is a key concern that helps inspire Koinonia's religious agrarianism and critique of empire); and that the community and its members would honor the covenant that they are "To be a steward of the earth, recognizing the beauty and wonder of the earth as God's creation" (Castle 2007).[10]

This last covenant is central to my argument for this book. First, notice it was added in 2002, well after the Age of Ecology of the 1970s, the publication of Lynn White's treatise, Earth Day, the Rio Earth Summit, and the slow realization that humans are dramatically altering the climate. In other words, popular understanding of human impacts on the environment had percolated throughout US society by 2002, and such concerns were powerfully felt even in Americus. Second, Koinonia is an intentional Christian farming community that is committed to stewardship of God's creation as one of its many principles and values. Furthermore, it is a Christian farming community that has a very strong critique of industrial farming and late capitalism. This combination has been a central feature of agrarianism in the United States dating back to the late 1800s.[11]

In April 2003 this covenant was further elaborated on during a Visioning Weekend. Fifty community members and friends attended this event and agreed to proactively meet five challenges Koinonia faced. The fifth challenge was the

"Demonstration of new community based and environmentally responsible ways of farming, including availability of plots to foster economic development for individuals and families" (Castle 2007). This dual covenant and concern, coupled with a history of progressive farming dating back to Jordan, inspired Koinonia to pursue permaculture and organic farming methods. This combination of principles and practices that are built on a unique institutional history make the community a leading exemplar of Christian religious agrarianism in the United States.

KOINONIA'S CURRENT STRUCTURE AND VISION

In November 2007 the community accepted a new vision statement that reads, "We are Christians called to live together in intentional community sharing a life of prayer, work, study, service and fellowship. We seek to embody peacemaking, sustainability, and radical sharing. While honoring people of all backgrounds and faiths, we strive to demonstrate the way of Jesus as an alternative to materialism, militarism and racism." Koinonia has established a community structure that developed in response to the turbulent 1990s and the visioning exercises in the early 2000s. The community selected three ideals to help structure and guide its next fifty years of existence that emerged during the visioning weekend and are education, hospitality, and sustainability. Five days a week the community has a shared lunch that is open to visitors and a community potluck on Sunday evenings (the earlier practice dates back to when the Jordans founded the community).

Besides these shared meals, the community has a museum with artifacts that outline Koinonia's rich history. There is a gift shop on site that sells various books authored by Clarence Jordan and a wide variety of community products. Visitors can buy T-shirts made from organic cotton; eggs, produce, and meat produced on the farm; fair-trade chocolate products; peace poles and Cotton Patch mugs made by community members; and a wide variety of books about sustainability and liberal, progressive Christianity. There is also on-site camping and a dormitory-style building where people can rent rooms and take part in community life for a short time.

The names of the rooms in the dormitory housing are a "who's who" of progressive Christian and non-Christian peacemakers, and each room includes a write-up on the person located on the door and bed stand. During my stay, names and write-ups included Martin Luther King Jr., Dietrich Bonhoeffer, Dorothy Day, Mother Teresa, Mahatma Gandhi, Rachel Corrie (a young US woman killed protesting the construction of Israeli settlements on Palestinian lands), Rosa Parks, and Nelson Mandela.

Two buildings over from the dorms and cafeteria[12] is a community library open to visitors which contains sections of books on crafts and building, peace, nonviolence, community living, family and marriage, natural health, and religion. In the section on science visitors can peruse books by Aldo Leopold, Charles Darwin, E. O. Wilson, Al Gore, and Helen Caldicott. In the section on gardening there are titles by Wendell Berry, a 1972 edition of Rodale Press's famous *Organic Farming*, a wide variety of books devoted to maintaining soil health via organic farming methods and composting, Albrecht Papers, and selections on the founder of biodynamic farming, Rudolph Steiner.

The community's outreach includes providing tours for school, church, and other groups, where upward of 2,000 visitors pass through the community in very busy months. The tours include a showing of *Briars in the Cottonpatch*, a video about the history of Koinonia, and a guided tour through the permaculture gardens. These activities are geared toward professed goals of hospitality, education, and sustainability and the tour content reflects the community's shared concerns for racial and gender equality, acceptance of immigrants, concern for the environment, and active support of social justice issues.

Those who want to go deeper into Koinonia's patterns and rhythms can apply for three- to four-month-long seasonal internships. Resident interns help with farming, work in the pecan business, share building maintenance duties, and complete office tasks. Those interns who are called to become community members can apply to become community partners and if accepted are able to reside at Koinonia. After some time (up to a year, sometimes more), the prospective member will either leave or apply to be a candidate for stewardship. Stewards are those who have (ideally) committed the remainder of their lives to residing at Koinonia and in so doing take part in its various ministries and requisite jobs and tasks. This system is intended to keep another scenario of embezzlement from occurring, protects the community from free-loaders, gives current stewards the opportunity to accept candidates for stewardship, and allows visitors and interns an opportunity to experience community life to see if they are truly committed to Koinonia.

In concluding his history of Koinonia, David Castle writes, "Prophetic organizations are always called to be the forerunners of the next era. . . . Dramatic environmental and milieu changes plus staggering technological developments require new ways of being and new ways of working. How to be relevant today as a 'demonstration plot for the Kingdom of God' is our current challenge. We are called again to play a decisive role in defining and shaping the course of human life on planet Earth" (2007). This calling has motivated Koinonia from its inception, as founders and participants have sought to be a demonstration plot for interracial living; to challenge "empire" and the military-industrial juggernaut; and to embody and put into practice sustainable farming techniques to demonstrate sustainable stewardship. In the pages and chapters that follow, I examine how Koinonia is currently embodying sustainable stewardship by putting into practice their religious agrarian calling as they hope to help influence the course of human–nature interactions so they are more just and sustainable. First I want to dig deeper into Koinonia's critique of empire and see how this critique is consistent with traditional agrarian critiques of industrialization. As we will see, this critique of empire partly motivates the community's sustainable farming practices.

NEW MONASTICISM AND SCHOOLS FOR CONVERSION

I return to one of my larger arguments about lived networks by quickly exploring Koinonia's continued history of actively partnering with other like-minded groups. These lived networks include past relationships (agricultural and religious) with Bruderhof, Hutterite, and Society of Brothers communities in the 1950s and a variety of current lived network relationships today.[13] These current lived networks include partnerships (whether by correspondence, physical visits, meetings at various conferences, invited presentations, or sharing links on websites) with progressive Christian social organizations Sojourners and Catholic Social Worker Houses. These are two major Christian groups that actively campaign for social and environmental justice. Those in Koinonia also maintain active lived network partnerships with communities that have been inspired by former members, like Habitat for Humanity and Jubilee Partners (the latter group houses immigrants who are taught English and other skills so they can adapt to US society). Other lived network relationships include active collaboration with sustainable agricultural networks such as permaculture groups, fair trade suppliers, and Georgia Organics (the state's most active and visible organic group that helps farms gain federal certification).

Finally, Koinonia maintains active membership in an emerging domestic Protestant movement called New Monasticism and its partnership and membership with New Monasticism enables Koinonia to further its own pacifist history. This is because both New Monastics and Koinonia share the critique of government-sanctioned military force and economic machinations that benefit corporations at the expense of the world's poor and the planet's ecosystems—for both groups, this symbiotic relationship is what constitutes "empire." Koinonia and New Monastics feel this is antithetical to the empire of love and compassion that Jesus preached and embodied.

The New Monastic movement is partly an anarcho-primitive challenge from those within Protestantism who believe that Christ's true teachings and actions require them to move to the margins of US society and minister to the poor, homeless, and broken. In his study of an emergent US monasticism, Charles Fracchia argues, "What I see developing as one current of the 'new monasticism' is not just an examination of and return to the ideals of the past, but a response to present society" (1979: 5). His study prefigures the response to the present waste, destruction, and spiritual anomie of North American society (including "fallen" Christianities) that the New Monastic movement feels called to challenge. He writes that, "A tendency towards the production of quality food . . . and towards the production of both aesthetic and practical goods that emphasize quality design and workmanship marks the growing cottage industries of the 'new monasticism.' Intimately related to this concept of jobs, work, and money is the concern for environment. Taking as little as possible from the environment is a corollary of the philosophy of working to sustain oneself" (ibid.: 19). It is possible to trace this same trajectory in Koinonia's history, from its back-to-the-land ethos of the 1960s and 1970s, increasing environmental concern from the 1980s onward, and desire for authentic Jesus-centered communal life. This institutional and communal history has resulted in today's community structure and Koinonia's active use of permaculture and membership in New Monastic networks.

New Monasticism has developed within the larger lineage of monasticism that Fracchia describes. This much more contemporary monastic movement has roots in Jonathan Wilson's 1997 treatise, *Living Faithfully in a Fragmented World: Lessons for the Church from MacIntyre's* After Virtue. At the time of publication, Wilson was associate professor of systematic theology and chair of the Department of Religious Studies at Westmont College in Santa Barbara, California. In his book, he laments that the current Western church is not living up to the life and teachings of Jesus and is not faithfully preaching the Gospel. In other words, he argues that those Christians who are faithfully incarnating the pre-Constantinian church's example have become a minority voice within a larger, secular, consumer, and thus compromised Christian country. Wilson believes Christianity is compromised because as US culture becomes fragmented along pluralistic lines, the church becomes fragmented. As he states, "Th[e] gospel and mission of the church never change, but the circumstances in which we witness to and live out the gospel do change. With changing circumstances comes the need to rethink how the church lives faithfully and witnesses the gospel" (1997: 2–3).

Wilson's analysis supports my argument for a lived network approach to understanding the greening of some North American religions—in this case, Protestant Christianity. The larger New Monastic movement takes Wilson's clarion call for reinhabiting Christian community (Wilson 1997: 65–66) to heart and accepts as valid his claim that "there are times—and I have argued that this is one of them—when the life of the church has been so compromised that we no longer are capable of fulfilling faithfully our mission. At such a time, the church must withdraw into a new monasticism, not in order to avoid a 'bad' society, but in order to recover faithful living and a renewed understanding of the church's mission" (ibid.: 71). Wilson ends his theological discourse by outlining four characteristics of a new monasticism (ibid.: 72–75), pointing out that "Although the new monasticism must be intended for the whole people of God, the discipline that it requires will be achieved only through small groups of disciples that are themselves committed to the vision and discipline . . . oriented around particular work or life circumstances" (ibid.: 74). For Koinonia, this vision and discipline are oriented around a particular lived embodiment of Christianity that dates back to 1942 and now includes a commitment to sustainable stewardship and religious agrarianism.

Wilson's heartfelt exegesis on reinhabiting the Christian church in a fragmenting world struck a chord with other Christians living in community. His call for radical new discipleship triggered a response from these less well known but vibrant fringes of Christianity in the United States. The New Monastic Library now publishes the series Resources for Radical Discipleship, and Jonathan Wilson-Hartgrove (not the same person as Jonathan Wilson) is their series editor. The series foreword reads,

> For over a millennium, if a Christian wanted to read theology, practice Christian spirituality, or study the Bible, she went to the monastery to do so. There, people who inhabited the tradition and prayed the prayers of the church also copied manuscripts and offered fresh reflections about living the gospel in a new time. Two thousand years after the birth of the church, there is a new monastic movement stirring in North America. In keeping with ancient tradition, new monastic's [sic] study the classics of Christian reflection and are beginning to offer some reflections for a new time. The New Monastic Library Series exists to share reflections from new monastic's [sic] and to print classic monastic resources unavailable elsewhere. (Rutba House 2005)

These New Monastic leaders further recognize their indebtedness to Wilson, writing, "Our origins are primarily in free-church Protestantism, and our communities do not often look much like traditional monasticism. We stole the term 'new monasticism' from our friend Jonathan Wilson and from his theological reflection on the work of Alasdair MacIntyre" (Stock, Otto, and Wilson-Hartgrove 2007: 1).

A variety of church and Christian intentional community leaders responded to Wilson's call by organizing a gathering at St. John's Parish Church in Durham, North Carolina, in 2004. This meeting was hosted by a New Monastic center, Rutba House, and the participants claimed that "In an age when 'Christian' America is the 'last remaining superpower' in an all-out 'war on terror,' we've begun to think that once again it is time for a new monasticism. Indeed, this is how we see the Spirit moving in North America today" (Rutba House 2005: ix–x).

To this end, the group developed twelve "marks" of the new monasticism they are helping birth on US soil. These are not meant to be comprehensive but "helps name the unique witness these neo-monastic communities have to offer the rest of the church" (Rutba House 2005: x). The twelve marks guide the lifeworlds of New Monastics and encourage them to

1. relocate to what they call abandoned places of empire, which include inner-city urbanscapes
2. share economic resources with the needy and other New Monastics
3. offer hospitality to strangers
4. fight racism within churches and communities and work for reconciliation
5. humbly submit to Christ's Body as it is embodied in the Church
6. form in the way of Christ and rule of the community
7. nurturing common life among the members
8. support celibate singles as well as married couples and families
9. live in proximity of other New Monastics who share a common approach to life
10. care for the part of God's Earth that is given to them to steward and care for local economies
11. commit to conflict resolution and peacemaking
12. live a life of disciplined contemplation. (ibid.)

The group justified these marks by asking, "Why wouldn't the church who finds her identity in a history of exodus and exile, in the story of a Refugee who was executed by an Empire and resurrected by the King of the universe, want to be where God has been active all along—on the margins and in the abandoned places?" (ibid.: viii).

In another book, Wilson-Hartgrove tellingly writes that,

> In every era God has raised up new monastic's [sic] to pledge their allegiance to God alone and remind the church of its true vocation. These people have not been perfect. . . . These saints who've called us back to our roots generation after generation remind us that the roots of God's kingdom are rhizomes. They spread beneath the surface, effecting change from below. It is a quiet revolution—one that is often ignored by the newspapers and missed by the historians. But it is, in the end, how God plans to save the world . . . radical hope is the heart of the monastic impulse. (2008: 54–55)

We see here Orsi's trope of lived religion in practice and that Koinonia is an active partner in this emerging new monastic network that has as one of its marks caring for the Earth.

These marks and this overarching sentiment help ground Koinonia into a larger neomonastic movement that is challenging today's "empire" of corporate-funded lobbying (*pace* the US Supreme Court ruling of *Citizens United*), military excursions, and neglect of the poor. Clarence Jordan can be interpreted as manifesting this spirit when he created Koinonia at the margins of abandoned farmland in rural Georgia during the height of the racist Jim Crow Southern empire. Marks two, three, and four were also put into practice by Jordan well over sixty years ago. Mark ten is significant for the purposes of my argument. Koinonia has active partnerships with a local coffee shop that sells organic and fair trade products, sells their produce at various local farmers' markets, and is in the process of creating a meat CSA, where people in the larger community can buy shares in the cows, pigs, and goats raised on the farm. Last, by creating a permaculture demonstration plot literally at the margins of today's corporate controlled, globalized, monoculture food system, Koinonia embodies a Christ-inspired New Monastic religious agrarian lifestyle.

Another New Monastic author, Tom Sine, writes that this movement has emerging leaders who are, "committed to calling people to a more authentic, embodied, whole-life faith . . . [who] tend to be outwardly focused in mission . . . are relational, organic and communal with virtually no bureaucratic, hierarchical models of leadership, unlike many denominational and nondenominational churches . . . [and] *tend to be concerned about a broad range of social issues, including social justice, reconciliation and creation care*" (2008: 39; emphasis added). Such concerns have led many New Monastics to create food pantries and soup kitchens at houses where they live in Christ-centered community. Living in community enables New Monastics to live lives of contemplation and dedication to peacemaking and the poor, serving those abandoned by empire without judgment. Many have also helped create community gardens and undertake other activist work in racially divided or economically depressed parts of the country.

This broad range of attributes and these issues of concern have been present at Koinonia since it was founded, and they have gained even more support in the ideals of today's community stewards. Community members are passionately committed to issues of social justice, reconciliation, and Creation Care. Furthermore, Koinonia regularly hosts Schools for Conversion workshops. These are events where the twelve marks are discussed and those who attend are given tools for putting radical discipleship into action in their own communities. Some of these tools are related to mark ten, which Norman Wirzba interprets is a mark about tending to the plot of God's Earth. Wirzba describes four possible New Monastic practices, including the practice of growing a garden. As he writes, "The palpable identification between humanity and soil (*adam* from *adamah*) is lost on us" (2005: 145) and growing a garden (preferably organic) can help overcome this lack of identifying with the Earth. By participating in and hosting Schools for Conversion, Koinonia maintains an active membership in lived networks of Christians who are motivated to care for the Earth by participating in sustainable farming and strive to embody environmental justice in today's wasteful industrial food systems.

THE AGRARIAN EXAMPLE OF KOINONIA

Koinonia's current farming practices and agricultural landscape took root largely due to the vision, dedication, and hard work of a community member who lived on the property from the late 1980s until February 1998—Jim Everett.[14] The Mennonite Church sent Everett to Koinonia after graduating from Michigan State University in 1983. For a year and a half, he commuted between the Habitat office in Americus and Koinonia, where he worked under the head gardener, Ray Rockwell.

A strong desire to travel overseas landed Everett in Bangladesh where the Mennonite Central Committee assigned him to live from 1985 until 1988. As he recalls, "Being rather disillusioned with such work by the end of it, and not having other ambitious visions, I decided to return to Koinonia as a volunteer in June 1988 and worked my way up through the membership process to the stage called 'partner' by 1990 or '91. This is similar in nature to what is now referred to as a 'steward' and I was the last partner to join the community as it was then structured (and also the last to leave, in Feb. 1998)."[15]

Jim Everett was present during the community turbulence of the 1990s, and his own recollections are that the years from 1993 until he left were the most trying times. As he explains, "Koinonia had been for many years racially divided between the white partners, to whom simple living was cutting-edge spiritual discipline, and a group of long-term black employees, with sharecropping in living memory and no desire to move toward anything resembling such a life." After the board of directors, which served as a legally mandated advisory role to the community partnership, switched the community to a corporate structure, the community decided to pay members based on comparable wages for similar jobs within the larger region. The board also hired its first CEO, "a charismatic black preacher who believed in prosperity theology . . . have faith and act as if you have it. The result was that within two years [1993–1995] most of the former partners had left, the place was some $800K in debt, and clear cutting of the forestland was under way to finance this. At the same time 'modernizations' were undertaken without regard to economy . . . carpeting and air-conditioning the office, removing most woodstoves and replacing them with gas heaters, etc." This is Everett's perspective on the beginning of the "dark night" of the community's soul that eventually ended in two cases of embezzlement by CEOs (Castle writes about these cases in his community history).

Within this milieu of changing corporate structure and conflicting community visions, Everett had successfully built up Koinonia's organic garden. The garden he worked in was part of a larger community history of organic farming as this term is commonly understood and defined in today's parlance. Everett writes that

> The tradition of organic gardening at Koinonia began in the early 1950's with Will Wittkamper, whom I never met, though I did get to know his wife before she died . . . she was still at Koinonia when I returned from Bangladesh. Will used to debate organic versus conventional farming with Clarence frequently, and the tradition of the two systems existing side by side continues to the present, presenting a unique opportunity for people to experience both (usually, I think, to the benefit of organic gardening at Koinonia and at large!).[16]

Although Jordan was progressive about certain farming practices, he was open to using chemicals when needed. This was in the 1940s and 1950s when such a perspective was influenced by the agricultural training Jordan received during World War II, when the use of chemical weapons was slowly being introduced to US farmlands. His use of emerging methods was geared toward helping the community and local sharecroppers achieve financial solvency, and within this context, strategic use of chemicals was allowed. As Everett relates, Jordan was open to debating the merits of this strategy. This tension exists to this day on the campus as most of Koinonia's income is generated by their pecan products, made from monoculture pecan groves that are treated with chemicals.

Everett notes that from Wittkampers's time until Everett's hard work, "the main goal of the gardens was food production for the people living on site, with occasional local marketing." However,

> By 1995 the organic garden was nearly the last vestige of the former community left, and it represented alternative values, liberal theology, and simple lifestyle; as over against the more mainstream values of the regime then in power. Moreover, each department of Koinonia was urged to become self-sufficient or more financially, and this prompted me to have the garden area certified organic and begin marketing produce through a cooperative then in existence in Atlanta.

Community infighting, politics, and financial demands eventually led Everett to leave the community and delve deeper into practicing, embodying, and promoting permaculture. Today he says he is "a happy pagan in the woods. The permaculture ethics [see chapter 6 for more on these ethics] are the closest thing I have to moral codes these days." He and his wife created their own permaculture homestead about one and a half hours away and actively returned to Koinonia to help lead various permaculture trainings and to work with its current lead farmer. This connection continued until the Everetts relocated to Oregon in 2010.

Jim Everett defines permaculture as "a system of overall farm design which includes organic, and even occasionally, aspects of conventional farming. Its goal is not to be dogmatic but to be effective. For instance, I still have some dumpster-diven chemical fertilizer. I have been gradually dribbling it out, mostly as a starter for the humanure compost" he and his wife were making at their homestead in the woods outside of Macon, Georgia. This definition highlights some of the internal tensions within sustainable agriculture enclaves and the polysemous meanings of sustainable agriculture. For Everett, permaculture can use chemicals if the need arises. Proper permaculture design and implementation should make this a last resort, for if a system is designed efficiently and properly, all the nutrient needs of the entire system (outside of sunlight and rain) should be generated by the system. This is an ideal that takes many years to achieve in practice (if ever, as there is no truly closed system in nature), and many permaculture farmers do use off-site inputs to keep the system running. Which inputs are permitted can thus become a point of contention, as some permaculturists are fine with chemicals, and others eschew these as they see them as being inimical to the sustainable ethos of permaculture.

Everett was exposed to ethnobotanical farming, permaculture, and agroforestry in Bangladesh, at which point he delved deeper into his love of plants and organic farming. His experience of forest-based farming, coupled with the edible forest ideal of permaculture, resulted in his efforts to design Koinonia's certified organic gardens along the tenets of permaculture. Both the current farming activities and sustainable agriculture vision of Koinonia thoroughly bear the stamp of his influence. Part of his legacy has manifested in a program of permaculture trainings, held once or twice a year, during which the community's land becomes a living classroom.[17]

Permaculture is rooted in the work of Australians Bill Mollison and David Holmgren.[18] Mollison defines permaculture as "a philosophy of working with, rather than against nature; of protracted and thoughtful observation rather than protracted and thoughtless labour; and of looking at plants and animals in all their functions, rather than treating any area as a single-product system" (Mollison 1979: 1). As Wes Jackson says, current industrial farming privileges part over whole, whereas sustainable farming methods like permaculture privilege whole over part (1987). This is one of the key critiques of modern farming by environmental agrarians and is a consistent theme found within Wendell Berry's corpus. Of import to my argument is that religious agrarians agree with this critique.

The fruits of this working with and observation is to create permanent agriculture based on systems of "whole landscapes and soil complexes" (Mollison 1979: 3) that climax in permanent, edible forests cultivated by "generations of care and knowledge . . . found in stable communities" (ibid.: 3). This matches well with the agrarian idea of cultural permanence in place, where nonindustrialized sustainable farming is practiced and passed down through successive generations.

As with agrarianism proper, permaculture is not a romantic or naive worship of nature and soil. Rather, it focuses on human needs and encourages disciplined action in nature that leads to farm health. This is seen in Mollison's influential early work when he writes, "If there is a single claim that I could make, in order to distinguish permaculture from other systems of agriculture . . . is that [it] is primarily a *consciously designed* agricultural system" (Mollison 1979: 6; emphasis in original). Mollison and Holmgren teach that an agricultural system that is consciously designed saves energy within itself; can handle outside energies from sun, wind, and fire without being overwhelmed; allows for the arrangement of plants so they form symbiotic relationships with other plants, so the system flourishes; allows for the system designer to place plants, houses, and other material artifacts in desirable locations; can be designed to suit a specific climate and a specific landscape/ecosystem; and can provide for energy, construction, recreation, and food needs (ibid.: 6). However, and this is key to my argument about religious agrarianism, the values that influence a permaculture designer's worldview will significantly determine the type of food system they attempt to create. Here we see that the values that permeate permaculture, including a holistic worldview and ethical sharing, are the same found in all forms of alternative agriculture and thus religious agrarianism.

Since Holmgren and Mollison founded permaculture in the late 1970s, a vibrant global network of workshops and trainings based on their initial insights has developed. The movement is particularly strong in Western Europe, North America, and Australia, with a US publication, *The Permaculture Activist*, being one of the leading voices within the movement. Koinonia strategically advertises its trainings in this magazine and thus further cultivates various lived networks in permaculture circles.[19] Moreover, Chuck Marsh, one of North America's permaculture pioneers who helped build and design Earthaven ecovillage in North Carolina, was invited to Koinonia to help develop their vision for sustainability in June 2008.[20] Thus, Koinonia's farming is thoroughly influenced by permaculture ideals, and the community is immersed within regional and national permaculture networks.

Koinonia brings a unique Christian perspective to permaculture, thereby helping mold their shifting, maturing, sustainable religious agrarianism. One community member who contributes to this and who moved to Koinonia with his wife in the mid-1990s is Patrick Smith. After living in community in California and working with dairy farms in the Northeast, the Smiths came to Koinonia to raise their family. Smith had taken a month-long wilderness survival class before their move and currently is an avid dumpster diver. He also wild crafts (harvests undomestiated plants and animals) various food sources for his family. Upon arriving at Koinonia, he worked closely with Everett before the latter left the community, leading Smith to explain that Everett's permaculture worldview greatly influenced his own about farming, as well as the perspectives of the current farm leader, Joel Taylor.

In April 2002 Smith helped organize a seminar hosted by Koinonia titled "'Celebrating God's Creation!' Exploring Demonstrations of Ecologically Sustainable Living." This seminar was part of a larger series of seminars titled "The Clarence Jordan Center for the Advancement of Christian Discipleship." The flyer for Smith's seminar states, "We will be involving participants in the Christian life of the current day Koinonia community as well as exploring other 'demonstration plots' of deep ecology within The God Movement. In continuity with Clarence Jordan's original vision, Koinonia's current residents and staff continue to struggle to discern the meaning of living sustainably in Christian community in the new millennium." Part of the objectives of the weeklong seminar was to "deepen the theological foundations upon which we can build a more socially and environmentally active church." It is possible to read in Smith's vision many of the same concerns shared by contemporary New Monastics. This seminar places Koinonia at the early vanguard of Christian communities that have been attempting to discern and put into practice sustainable lifeways. The institutional support Smith received in developing and hosting this conference shows Koinonia's strong and continually maturing commitment to sustainability, especially via the vehicle of agriculture. Readers should note that this support includes an important theological dimension so that Koinonia is also in the vanguard of religious agrarian communities.

Smith scheduled the seminar to occur during the annual Earth Day celebration in April. Over the course of the week, seminars included some of the following workshops, as described in the seminar pamphlet: Opening the Book of Nature, a workshop "designed to help you connect to the Christian tradition of learning spiritual lessons from nature"; Religion, Theology, and Environmental Crisis; a talk by Everett titled "'Spiritual Permaculture': Redemptive Living on Sacred Land"; and a wild edibles walk led by Everett and Smith, called "Seeking the Morning Manna." Last, workshops were offered on ecofeminism and "Christian Activism in the Environmental Peace Movement." Such topics articulate the diverse approaches that Koinonia and others in progressive Christian circles are developing in regard to interrelated issues of peace, activism, justice, and sustainability. These approaches and theological standpoints thoroughly infuse and influence Koinonia's approach to religious agrarianism.

Although the conference was open to people of all religions (or none at all), it largely attracted other progressive Christians. Presenters included Quakers, Presbyterians, Congregationalists, and representatives from Habitat for Humanity, Oak Ridge Environmental Peace Alliance, and Servants in Faith and Technology—Southern Institute for Appropriate Technology. Again we see that Koinonia has developed active lived network partnerships with other groups and communities of faith in regard to sustainability. This unique fusing of permaculture principles, lived networks, and Christian practice makes Koinonia an exemplar of religious agrarianism and situates the community at the forefront of a rapidly greening North American religious landscape that is taking food issues seriously.

The evolution of Patrick Smith's perspective on Koinonia's religious agrarian practices can be seen in an article he wrote for the quarterly newspaper that gave a summary of the group's farming history.[21] Tasks and jobs are divided into separate categories under the new community model, and each task has a coordinator. Farming has its own team, which is typically staffed with interns, and a coordinator makes most of the important decisions regarding how the farm will follow the covenantal vision worked out earlier in the new partner covenant. Smith writes that the new coordinator "has fully entered into his calling to restore the visionary work began by those who dressed and caressed the trees and soil before him. I am convinced that Joel Taylor is the one who is now called by God to lead us into a renewed commitment to care for the environment which sustains us" (*Koinonia Farm Quarterly* 2009: 5).[22] Furthermore, Smith finds evidence that "From the very beginning of Koinonia, there has been recognition of how important a responsibility it is to be good stewards of this corner of creation that God has entrusted to us" (2009: 5). Such a clarion call by a community member that recognizes the sacredness of creation is historically abnormal, in the US South and in Christianity broadly. Yet it is precisely such a view that permeates and legitimates religious agrarianism. This call and others like it mark a unique development in religion–nature–food interactions.

A deeper analysis of specific religious agrarian practices currently being implemented by Koinonia's farmers and an analysis of interviews with some of the community members in charge of implementing these practices occurs in chapters 4 through 6. I now turn my attention to a summary of Christian approaches to agriculture, especially sustainable agriculture. This turn helps provide more context for what is unique about Koinonia and shows how they fit within larger Christian movements and views that advocate for sustainability.

A BRIEF HISTORY OF CHRISTIAN FARMING

Because of the conversion of Emperor Constantine, the history of Christian farming is more complex and widespread than the history of Jewish farming

I explore in the next chapter.[23] This aspect is due to gross numbers and total acreage, given that Christianity became the religion of empire. Its dominance and hegemony was challenged by internal division and the rise of Persian and Ottoman Muslim empires, but given the overall time period and span of continents, it is only possible to give broad strokes of a historical outline of Christian farming as Christianity spread beyond early urban enclaves. These strokes occur within the realization that there never has been a monolithic or internally consistent Christianity, either. In this quick summary we must also be cautious, for "The single most important generalization to be made about the attitudes toward nature held by the peoples of the classical world is that these varied greatly throughout the long span of ancient history" (Glacken 1967: 13).

The various early Christian communities struggled with their identity—were they Jews and thus an agricultural people now living in diaspora, or were they a new people? The New Testament author now known as Paul helped clarify this ambiguity in forceful terms, so that the agricultural history of the Hebrew Bible and the sacrifices to Yahweh quickly lost their power as identifiers within emerging proto-Christian communities. Christ's atoning sacrifice was for all people, and despite three centuries of persecution, this message nonetheless found a home in Constantine and then the Roman Empire. The centralized roads and agricultural practices and technologies and overall politics of Rome became Christianized, while Christianity became Hellenized.

Underneath this Hellenistic bedrock was a split between rural and urban, already evident in the philosophy of Socrates. Although Aristotle developed a natural philosophy, the Greek antipathy toward the rural and toward nature generally was influenced by the thoughts of Socrates and Plato (Bender 2003: 162–66) as well as by the utilitarian and aesthetic views of nature found in the Stoics (Glacken 1967: 52). All told, the Hellenistic world was built on a cattle economy that led to massive deforestation, coupled with the glorification of the rational, urban world that clung to the idea of a past golden age of soil fertility (ibid.: 132–33). However, the Hellenic concept of a virtuous citizenry was influenced by Aristotle's *Politics*, including the idea that "a healthy agriculture [is] an indispensable condition of a healthy society" (Montmarquet 1989: 26).

On this mix of Israelite and Hellenic worlds, Christianity took root. The cosmology in Genesis became the lens through which Christians viewed nature, with this view being supreme until the beginning of geography in the 1800s and then Darwin's theory of evolution in 1859. The story of the Fall from Eden in Genesis "became important to the Christian idea of nature because it is the source of the belief, widely held through the seventeenth century, that the fall of man has caused disorder in nature and a decline in its powers"

(Glacken 1967: 154). Such views were wedded to scriptural insights and teachings about famines and failed crops, suggesting that "some form of recognition existed about the fragility of agricultural production, although in an era without meteorology, the understanding for crop failures was related to concepts of sin and lack of fidelity to covenants" (LeVasseur 2014b).

Another development of the fusion of the Genesis cosmology and Hellenism is that "The theme that man, sinful though he be, occupies a position on earth comparable to that of God in the universe, as a personal possession, a realm of stewardship, has been one of the key ideas in the religious and philosophical thought of Western civilization regarding man's place in nature" (LeVasseur 2014b: 155). This idea of stewardship, and the cosmology offered by Genesis in general has been decried by those like Lynn White Jr. and Norm Habel, who blame religion (especially Christianity) for the current ecocrisis (although both point to the second Genesis creation story as being amenable with stewardship); while apologists like Calvin Dewitt and other Christian ecotheologians interpret this passage as a challenge to be benign stewards. The latter reading is one that members of Koinonia (and equally some Jewish religious agrarians) employ to justify their religious agrarianism (on this, see especially chapter 4).

Agriculture and general human conceptions of nature in the Christian Middle Ages were built directly on a biblical worldview. Leading theologians like St. Basil and St. Augustine saw creation as an act of the Creator, so creation was viewed as an act of love. Although it is a place of sin and the goal is salvation, nature is something to be studied and appreciated, with Augustine even teaching that nature is not evil, humans are. However, "It is a distinction that lies at the root of Christian belief and in the Christian attitude toward nature: one should never become so entranced with the beauties of nature that he mistakes them for anything other than creations like himself" (LeVasseur 2014b: 197). Although there was a physico-theology that developed during the patristic period, there was also the development of an ascetic, monastic, otherworldly view within Christianity that included feelings of contempt for nature yet saw manual labor as a source of pleasure.

Under St. Bernard, St. Francis, and Alan of Lille, monastic communities developed and believed that through their labor it was possible to create paradise out of chaotic wilderness. The result of this worldview was the clearing of forests throughout Europe. At the same time, Muslim ideas about nature crept into the Christian West, building on Aristotelian ideas of telos so that by the time of St. Albert, ideas of a designed Earth influenced viticulture, horticulture, and agriculture. As the Catholic Church spread, physically and ideologically/theologically, especially after the Black Death (Thirsk 1997), agricultural land use patterns changed, as well: "Under the influence of the Roman conquest and the Catholic Church, vineyards took over land in all of Europe during antiquity" (Mazoyer and Roudart 2006: 276). Later in the Middle Ages, "the combination of technical skill in agriculture with the ability to provide spiritual care made the monasteries following the Benedictine Rule very powerful in many parts of Europe" (Glacken 1967: 308). Overall the positive valuation of labor, the view of humankind as agent of God working to create a Christian civilization that guided souls toward salvation, and increasing technological mastery over nature contributed to a reworking of landscapes across Europe with monasteries at the forefront.

Given the Church's power and wealth, "The Church benefited from a large part of society's surplus in this time period and used it to build lavish monuments, dedicated to the glory of God" (Mazoyer and Roudart 2006: 301). Furthermore, given the lack of a printing press and coupled with intellectual hegemony held in select circles, "A particular group of manuscripts from the early Middle Ages transmitted classical knowledge" (Brunner 1995: 33), including about nature, technology, and agriculture. This system of knowledge transfer means that "The continuity of Christianity played its part in the technological transfer of agricultural knowledge" (ibid.: 33). The cumulative result of the above processes was the feudal, agricultural Christian kingdoms that dominated the region until the 1600s, whence began the rise of mercantile capitalism and the onset of the Protestant Reformation.[24]

While this history is presented as linear, it was not without theological battles that influenced human–nature perceptions and interactions. The spread of Christianity was clearly met with resistance, and there were plagues, famines, floods, and droughts, so the common peasant working agricultural fields did not necessarily occupy a world of abundance. Furthermore, as environmental historian Elspeth Whitney observes about technological use in the Middle Ages, especially in agricultural contexts, "a variety of geographical, social, and economic factors and local and regional differences came into play that helped determine if and how technological innovations were adopted. The picture that has emerged of technological change in northern Europe between 900 and 1300 . . . is one of incremental change, diversity, and long-term diffusion combined with occasional instances of success and relatively quick innovation" (2013: 323).

It should also be kept in mind that "the productivity of [peasant] labor was fundamental to the vitality of the [medieval] economy as a whole" (Campbell 1995: 77). This is because domestic agriculture supplied the raw materials medieval industry needed and supplied towns with their fuel and provision needs.

With the onset of animal-drawn plowing and cultivation (as compared with use of the ard), coupled with fallowing, the use of manure, and the creation of hay fields to feed animal labor, Christian European countries in the north were able to expand production (Mazoyer and Roudart 2006: 259–61).[25] This production led to new landscapes, in part dominated by hay fields, and the excess food helped fuel the growth of urban areas and the prestige and wealth of the Church. This is seen in the creation of larger and new cathedrals, which would not have been possible without new agricultural technologies.

In other words, the Christian West (privileged because it is from this heritage that European colonizers and farmers of the Americas descended) was built on the backs of peasant agriculture. More so, "it is only with the central Middle Ages, from the eleventh to the thirteenth century,[26] that the agrarian systems characterized by the combination of fallowing and cultivation with the animal-drawn plow were widely developed in the northern half of Europe. Subsequently, they were transferred through European colonization [so that the] agricultural revolution of the Middle Ages carried the rural economy in the West to the threshold of modern times" (Mazoyer and Roudart 2006: 260). Almost all immigrants to North America were Christian agriculturalists, either directly or a few generations removed (except for mercantile capitalists, seamen, and those with city-based trades), and most brought with them this system of farming. The view of agricultural technologies and associated farming practices these largely Christian immigrants brought with them affected landscapes in the United States, including the farming options available to a young Clarence Jordan as he started Koinonia.[27] Furthermore, most immigrants were intimately familiar with agricultural cycles and lifestyles, as the urban–rural split of industrial agriculture is an invention built on a modern ideology and the abundance of cheap oil and natural gas.

The theological gestalt of Christian Europe, during its agricultural heyday and then under the burgeoning auspices of the Industrial Revolution, was that "Piety was an active ally compatible with . . . desired changes in nature. Creating a landscape fit for Christian settlement for conversion and colonizing was a reward of piety" (Glacken 1967: 293; see also Weber 2003). This worldview led to a "conquest of paradise," as European colonizers disrupted the New World with invasive flora and fauna and brought with them farming practices (and underlying conceptions of human–nature relations) that radically altered the landscapes of the continent (Radding 2005; Sale 2006).

This alteration of landscape is seen in early colonial enterprises based on plantation agriculture from Virginia south with this model supported by the Anglican Church (Isaac 1999); Anabaptist farming techniques brought from Europe and put into practice throughout the Midwest (Séguy 1973); and a subsequent industrialization of US farming practices (Hurt 1994). Within this worldview the technological developments of World War II,—including warfare chemicals that have since been used in agriculture, the discovery of oil and natural gas as sources of energy, and the move toward suburban lifestyles—all occurred. Their combination has led to the industrialization of agriculture and, importantly, the current rebuttal of this mode of food production by Wendell Berry and other ecological agrarians.

Finally, within this milieu Christian religious agrarians now operate, including those living and farming at Koinonia. If such a milieu is indeed vibrant and growing, then we should expect to see certain crystallized views that motivate Christian religious agrarianism. A key question becomes: what sort of ethics, both theological and environmental, are Christian agrarians using to inform their practices and critique of industrial agriculture? With the background of Western and North American Christian farming behind us, let us move on to answer this key question about motivating ethics.

ECOLOGICAL ETHICS, SUSTAINABILITY, AND CHRISTIANITY

> "I believe our Christian responsibility as farmers begins with the dirt. If God created the Earth, he must have put the fertility in the soil, and He surely intended for us farmers to try to keep it there for the sake of those who come after us." (Paul)
>
> "Maybe it isn't fair to ask the farmer to be Christian unless the rest of us are willing to be Christian too." (Carleton)
>
> "I never realized how complicated being a Christian in your vocation might get to be if you really took it seriously." (Carleton)
>
> —Excerpts from a round-table interview on "Farming as a Christian Vocation"(Greene 1949)

In this section I briefly explore emerging holistic human–nature concepts from within Western (especially US) Christian circles to highlight how Christian religious agrarianism in the United States is a recent phenomenon made possible by developments in ecotheology. I provide a few brief examples of US

Christians who are practicing forms of religious agrarianism similar to those practiced at Koinonia. These examples suggest that Christian religious agrarianism is a recent North American phenomenon in its nascent stages. Given growing ecotheological sophistication, the rise of Creation Care and concepts of stewardship, and climate destabilization, I predict this phenomenon is going to grow, with similarities and differences among the various groups extant in the United States (and Canada where relevant).

I begin by visiting the work of one of the leading ecotheological voices for developing a Christian ecological ethic, the recently deceased James Nash. Nash eloquently speaks about ecological sin and defines this as "a defiant disrespect or a deficient respect for the interdependent relationships of all creatures and their environments established in the covenant of creation, and an anthropocentric abuse of what God has made for frugal use" (1992: 119). By taking ecology and sustainability seriously, he argues that Christianity must redefine itself and reconfigure human–nature relations, which will lead to the redemption of humans from living in ecological sin. The path of ecological salvation rests on what Nash calls ecological dimensions of love, which are based on the dimensions of Christian love: beneficence, other-esteem, receptivity, humility, understanding, communion, and justice (ibid.). These dimensions of Christian love are on display at Koinonia, and many of them directly motivate the Christian agrarian practices of the community. Note, too, that the justice component of love is one of the key themes present in religious agrarianism, broadly.

Another leading voice within Protestant environmentalist circles is Calvin DeWitt, a zoologist who founded the Au Sable Institute. DeWitt begins a treatise on his hoped-for Christian response to environmental issues (2007) with a quote from John Calvin. Written in 1554, Calvin offers commentary on Genesis 2:15 and writes,

> The custody of the garden was given in charge to Adam, to show that we possess the things which God has committed to our hands, on the condition that, being content with the frugal and moderate use of them, we should take care of what shall remain. Let him who possesses a field, so partake of its yearly fruits, that he may not suffer the ground to be inured by his negligence, but let him endeavor to hand it down to posterity as he received it, or even better cultivated . . . let everyone regard himself as the steward of God in all things which he possesses. Then he will neither conduct himself dissolutely, nor corrupt by abuse those things which God requires to be preserved. (quoted in DeWitt 2007: 5)

This passage is significant because it evidences a Reformation-era concept of responsible, holistic, Christian-centered agriculture and because DeWitt strategically uses it to show other Christians that the great theologians of their tradition were concerned about proper, healthy human–Earth relations. In other words, if a luminary such as Calvin was concerned with "sustainable" farming, is such a concern no less important for Christians today?

DeWitt answers in the affirmative, arguing that the Creator offers seven provisions to humans so we may have a bountiful life. The second of these is soil building, where "The dynamic fabric of roots, soil organisms, and soils that bind together the surface of the biosphere makes one stand in awe of God's patience as Provider . . . Soil building helps to hold the whole world together. It helps support creation's integrity by renewing the face of the earth" (DeWitt 2007: 17). We see an ecologically informed understanding of the importance of soil in maintaining sustainable societies, a view assiduously championed by ecological agrarians and now religious agrarians. In a move similar to those made by ecological agrarians, DeWitt laments that "Agri*culture* is being displaced by agri*business*" (ibid.: 37). Such a view from Christianity is a recent development of the past ten to thirty years, made possible by the analyses of people like Wendell Berry, Wes Jackson, and Paul Thompson and organic luminaries like Howard and Rodale.

Even earlier evidence for placing what I call a religious agrarian concern for sustainable agriculture in US Christianity can be found in an edited volume titled *Farming the Lord's Land: Christian Perspectives on American Agriculture* (Lutz 1980). For example, in this volume, C. Dean Freudenberger, an agronomist and professor in the School of Theology at Claremont, writes that agriculture must be sustainable and just. He offers five suggestions that can move the nation toward a sustainable agriculture: a science of limits, an understanding of the social and environmental impact of farming methods, the adoption of biointensive agriculture based on crop rotation and biodiversity, smaller farm sizes, and just international food relations (1980). Freudenberger shares a profoundly religious agrarian concept at the end of this list, one that presages and affirms the motivating values that inspire the Christian agrarians (and New Monastics) of today. He writes, "Land is a gift in covenant. It is God's creation, not ours. It is a corporate gift for all generations. As Christians who are trustees of this endowment, we have freely accepted the invitation to achieve justice and sustainability in the management of the land and its water resources" (ibid.: 141). Such a sentiment thoroughly imbues the religious agrarian worldview of many of the Christians I researched and encountered at Koinonia. This is seen, for example, in the writing and conference organizing of Patrick Smith.

Given the back-to-the-land movements of the 1960s and 1970s, coupled with the Jesus people movement of the same time, it is certainly plausible and highly likely that similar Christian-based concepts of holistic farming and land stewardship were already present in North America at the time of Freudenberger's writing. The perceived evidence of the deleterious effects of industrial agriculture coupled with the continuing ecocrisis has led Christian attention to environmental issues to grow by leaps and bounds since the Rio Earth Summit in 1989. This growth has resulted in eloquent and sophisticated theological treatises on topics from sustainable agriculture (Graham 2005; Sutterfield 2013) to climate care (Spencer et al. 2009; Wilkinson 2012), to looking at evolution as evidence of God's grace and love (Johnson 2014) to engaging nature as a form of Christian spiritual practice (Chase 2011), to a whole gamut of Christian perspectives on a variety of interconnected environmental issues (Bingham 2009; McFague 2013; Van Wieren 2013).

With the onset of ecotheology (Ruether 1978), religious environmentalism, and the flourishing of a US sustainable food culture that is penetrating far into the mainstream, Christian communities the country over are rapidly developing religious agrarian values. This is resulting in a plethora of Christian religious agrarianism being put into lived practice around the United States today. Antecedents exist in the Shakers and in the long and continued history of the Catholic social worker movement, with movement houses often having organic gardens (as seen, for example, at Haley House in Boston). However, it is predominantly the past ten to fifteen years that have seen a proliferation of Christian communities and institutions engaging in sustainable farming, from Anathoth Community Garden in North Carolina (Bahnson 2006) to Lamb of God farm in Illinois (Moll 2007) to backyard church gardens increasingly sprouting up across the country (Rossi 2008) to helping the economically disadvantaged access healthy food by providing financial services (Sutterfield 2008). There are countless other examples to be found, and Koinonia is situated within this larger North American Christian religious agrarian milieu. Because they offer permaculture workshops, participate in a variety of lived networks, and host Schools for Conversion, Koinonia is a leading voice and exemplar of a wide variety of domestic US Christian groups attempting to put religious agrarian values into practice.

Beginning with chapter 4, we will come face to face with Koinonia's religious agrarians by hearing their voices and learning about specific farming practices they embody as they attempt to honor God's creation. These insights will be scaffolded on some of the insights shared here, helping create a snapshot of contemporary Christian religious agrarianism in the United States with

Koinonia functioning as the entryway. Before that, however, I introduce Hazon and Congregation Shearith Israel in chapter 3. With this introduction in place, a rich comparison between contemporary Jewish and Christian religious agrarian practices will emerge in chapters 4 through 6, where I begin the analysis of my fieldwork, arguing that the practices of Koinonia and Hazon reflect the greening of North American religions. As will be seen, their practices further evidence the integration of sustainable agrarian ideals and religious agrarian values of locality, health, and justice.

CHAPTER THREE

HAZON AND JEWISH RELIGIOUS AGRARIANISM

> Even though our ancestors didn't know this, and wouldn't have said it this way, our tradition compels us, mitzvoth, obligations, the Torah compels us to be sustainable in our agriculture. It compels us to take care of farmers and to not put small farms out of business . . . I think that there will be a growing trend towards ethics, environmentalism, ecology; eco-Judaism is going to continue to grow, continue to be influential on the Jewish community. Tremendously influential, I think. More than most people recognize today.
>
> —Rabbi Joshua Hillel, Congregation Shearith Israel

THE SPIRIT OF ADAMAH

My goal in this chapter is to introduce the larger context out of which religious agrarianism is developing. The specific focus here is Judaism in the United States, offered with the same caveat as my exploration of Christianity in the US: there is no singular, normative Jewish identity or position on anything, let alone food (see chapter 3, note 1). I present a brief overview of the role American Jews have played in US politics. This overview helps further one of my larger arguments about these leading edges of certain politically active US religious groups in regard to sustainable food issues and their potential to affect agricultural policies and (at least perceived to be) environmentally friendly lifestyle choices. I turn to an exploration of Judaism and ecology, briefly outlining some of the key Jewish approaches, both textual and organizational, that have emerged from within Judaism to environmental problems.[1] Couched within this analysis will be a brief history of Jewish agricultural farming communities that have existed in the United States. After this I discuss the more environmentally progressive manifestations of US Judaism, paying particular attention to Hazon. With this in place, I conclude by turning my attention to my key case study, Congregation Shearith Israel, providing evidence for why I believe their members are generating religious agrarianism-in-action.

The reader should keep in mind that, consistent with my lived network approach, what people do in the name of religion adapts and changes on the ground to meet the needs of practitioners. Of import is that this adaptation and change is influenced by membership in networks. As environmental agrarian concerns create hybrid spaces within religious communities, it should be understood that these communities will eventually change various aspects of their respective religious traditions. As this occurs, they will actively form and/or join lived networks devoted to sustainable food issues. Judaism in particular has a long history of flexibility (and counterhistory of orthodoxy) of interpretation extending back to the rabbinical codification of Halakhic law codes. For example, Burton Visotzky articulates, "This spirit of continuing change in Jewish law [via] Oral Torah [means the creation of Judaism is] an ongoing and not static process" (1999: 76). Evidently this process of continual change within Jewish lifeways and teachings is now responding to sustainable farming issues as Jewish communities redefine what it means to keep kosher; install organic garden plots in backyards, Jewish Community Centers, and synagogues; and reinterpret legal and rabbinic tradition in light of contemporary environmental sciences as these apply to agricultural issues.

POLITICS AND JUDAISM IN THE UNITED STATES

US Jews are disproportionately politically active compared to almost all other faith-based segments of the republic. Political scientists Kenneth Wald and Allison Calhoun-Brown posit that this is a result of the social status of Jews, who as a group face possible persecution and anti-Semitic policies and behaviors, and this sense of social marginality influences their political activism.[2] Furthermore, Jewish political activism in the United States tends to support liberal politics, even though most Jews are Caucasian and wealthy, two indicators that over the past few election cycles signify conservative, Republican voting trends. Despite these demographics, "many Jews actually regard liberalism as the very essence of Judaism itself" (Wald and Calhoun-Brown 2007: 274).

Wald and Calhoun-Brown provide other insights that provide explanatory power for my overall argument about religiously motivated lived networks of progressive environmental activism devoted to sustainable agriculture and larger agrarian issues. As political scientists, their methodology uses statistics, polls, voting records, and survey results to better understand voting behavior and political activities of religious citizens. One of their findings is that religious activists tend to focus their energy on public policy issues. This activism tends to be motivated by a mix of group identity, group status, theology, worldview, and institutional interests (although religious leaders tend to embody the latter). As I argue throughout this book, values, especially sustainable agrarian values, are shaped by and embedded within religious groups and networks. Thus, it is important to recognize that liberal Judaism in the United States has within itself a high level of grassroots religious activism that according to Wald and Calhoun-Brown is dedicated toward influencing public policy. This activism is readily seen in the organizational structure and political activities of Hazon, which range from bike rides for raising awareness about environmental issues like climate destabilization to supporting community-supported agriculture (CSA) partnerships.

Furthermore, regarding activism, including religious activism more broadly, Wald and Calhoun-Brown's work suggests that local community activists "are harnessing the potential of religious faith to organize people to address the material realities of their lives. . . . Working through congregations, these organizers use religious practices, worldview, and culture to empower people to address the concerns of their communities" (2007: 137–38). This sort of organization to address community concerns (and for the purpose of religious agrarian concerns, this community is extended especially to the soil—see chapter 4 for more) is evident in the local practices of New Monastics like Koinonia, and in the food campaigning undertaken by leaders within Hazon. An example of the latter is addressing material issues of food poverty (nutritional, environmental, and ethical) by creating lived networks that join CSA farmers with synagogues and temples around the country while advocating for Jewish CSA members to become politically active in local food politics. It is helpful to understand the progressive Jewish food networks as exemplified by Hazon in light of the political analysis of religious activism offered by Wald and Calhoun-Brown.

There is more to Jewish liberalism in the United States than simply fear of possible anti-Semitic, conservative voting regimes. Rabbi Sidney Schwarz, founder of the Institute for Jewish Leadership and Values, delineates a strong tradition of liberal, ethical concerns within the history of US Judaism. As he explains, "Few commitments unite Jews as much as a commitment to create a more just and equitable society" (Schwarz 2006: xxi). He recognizes a tension within US Jews along the lines studied by Wald and Calhoun-Brown. He calls the Jewish fear of persecution "Exodus consciousness," which is a trend within history of Jews to solely be concerned about their survival as a persecuted and chosen people to the detriment of engaging society at large and working on social justice issues for all. At the other pole is the Jewish capacity to generate, hold, and embody what Schwarz calls "Sinai consciousness." This is a definable trend within Jewish history to develop and work toward universal and altruistic ideals that apply to all people, regardless of religious identity and nationality.

Schwarz identifies a key biblical injunction that defines the purpose of Judaism and inspires Sinai consciousness. This is found in Genesis 18:19, when Yahweh tells Abraham to *la'asot tzedakah umishpat*, which in English can be translated into the mandate to "extend the boundaries of righteousness and justice in the world" (Schwarz 2006: 4–5). Furthermore, Schwarz's exploration of Jewish concerns for social justice includes a strident critique of US affluence and consumerism in solidarity with similar critiques levied by both New Monastics and Robert Bellah. This critique and the mandate to do what is right and just is reflected in the liberal voting behaviors and active social justice lobbying of many Jews in the United States.

Schwarz identifies seven core values of the rabbinic tradition that are related to social justice and have influenced the liberal political activism of US Jews (2006: chapter 8). Some of these include *chesed*, loving-kindness and compassion; *kavod habriot*, recognizing the dignity of all creatures because they are created by God; *bakesh shalom*, to seek peace and pursue it; and *emet*, inner and outer integrity and truth. A variety of these social justice concerns blend into Jewish environmental concerns, including loving-kindness and compassion (toward farm workers, farmers, and nonhuman life forms on farms); recognizing the dignity of all life so that traditional concepts of kosher are being challenged (see especially chapters 5 and 6); and the effort to match inner values with outer practice so that it is possible to live a life of integrity in regard to religious agrarian issues.

Schwarz offers a further insight into Jewish culture and politics in the United States, as well as an ambitious claim about Jewish history: "Judaism brought into the world the notion of social responsibility" (2006: 249). This history of concern for social justice and social responsibility now manifests in Jewish sustainable food circles, with a very clear mixing with quintessential agrarian concerns (as seen in the epigraph of this chapter).

JUDAISM, ECOLOGY, AND RELATIONSHIPS WITH THE LAND

Here I outline in broad strokes various Jewish relationships—material and ethical/religious/legal—with the land, placing more emphasis on recent historical manifestations found within a distinct North American and specifically US context.[3] This background forms a frame from which we can better understand the emergence of Hazon and Jewish religious agrarianism broadly. To understand modern, especially US Jewish ideas of and relationships with the environment, it is important to have a brief overview of Jewish history. Because early Israelites were a farming people, agriculture played a significant role in their annual ritual calendar. Many commentators note the agrarian basis and agricultural meanings of a variety of prayers and mandated practices found within the Hebrew Bible and early Israelite religion. In general, food played a large role in early Hebrew religion as well as in structuring social relations, including those with Yahweh.[4]

Jewish relations to the land are shaped by the fact that Jews have largely had to define their identity within minority enclaves amid and among violent anti-Semitic settings.[5] For example, Jews in Byzantium suffered under oppression from early Christians, who had turned their own rage of being oppressed by Rome onto their neighbors. Such rage resulted in growing oppression from the fourth to seventh centuries CE in Palestine, where Jews could not intermarry or own slaves.

These arrangements of empire precipitated "a first step in the alienation of Jews from the land, a process that ended with their becoming a nearly completely urban people in the Middle Ages" (Scheindlin 1998: 65). Once they became a thoroughly urbanized people living in ghettos, Jews (by this time it was rabbinic Judaism within diasporic communities) turned inward to the Written and Oral Torah. This overall historical process of turning away from an agrarian past (Davis 2009) was exacerbated when Arabs conquered Jerusalem in the eighth century. As Scheindlin explains, "Conditions under Islamic rule favored urban life and trade, and Jews were affected by this empire-wide trend, becoming ever more urbanized and drawn away from agriculture and toward commerce" (1998: 76).

A third historical arrangement—the lack of land titles—further removed the majority of Jews from agricultural livelihoods, customs, and lifeways. In medieval Europe, the great majority of Jews living under feudal arrangements "either never came into possession of land or were sooner or later driven from it" (Scheindlin 1998: 100). Living in such a context would have made it difficult for European Jews to develop in-depth agricultural practices, identities, and customs, so the idea of place would have retreated to the periphery of Jewish concern. In part the contemporary secular and then religious move toward Zionism, including a strong love of the land of Israel and urge to farm that land for self-sufficiency and Jewish identity, has been shaped by this history of displacement from Middle Eastern, North African, and European farmlands.

This brief overview does not capture the intricate complexities, politics, and counterexamples that equally constitute Jewish history. Nonetheless, and at a scale that far exceeds the historical relationship Christians and Muslims have had with the land and agriculture, Jews were systematically kept from developing a similar political, ethical, and living religious relationship with various landscapes. This history shaped Jewish religion, Zionism, and the kibbutz movement of the 1900s and therefore contemporary Jewish environmental concerns. Furthermore, it has directly contributed to the reason North American Jews are the most active globally in developing Jewish environmental ethics.

Jews in the United States meld scholarly, rabbinical, and lay approaches to the ecologic crisis within a larger US environmentalist milieu built on the legacies of Aldo Leopold, Rachel Carson, Earth Day, back-to-the-land, and environmental justice movements. As the leading Jewish environmental campaigner of Israel Jeremy Benstein writes, "the new creation of Jewish environmentalism is largely a product of North American Judaism" (2006: 211).[6] Jews in the United States, and those from around the world, are finding rich resources within this topophilically (Fu-Tuan 1974) disenfranchised history that nonetheless speaks to an ongoing dialogue of Jews with their God regarding eco-mitzvoth. Although such a move is a reinterpretation of tradition based on religious environmentalist concerns and not a traditional interpretation of Jewish religion, it is nonetheless a valid move and gives evidence to how religions morph and change based on the lived needs of followers (McCutcheon 1997; Mack 2001; Tomalin 2002; Gross 2003; Vasquez 2011).[7]

JEWISH AGRARIANISM IN THE UNITED STATES

Despite the foregoing history, there were indeed Jewish farming enclaves in Europe, Russia, and the Middle East.[8] However, most Jews in North America emigrated from urban Ashkenazi regions of Europe. Once in the United States, Jews of all cultural backgrounds settled in cities and spread westward. As they spread, they created numerous intentional Jewish communal agricultural experiments beginning over 150 years ago, with these experiments existing at various sites throughout the United States (Bartelt 1997).

The first Jewish agricultural colony in the United States was Ararat, founded in 1820; the peak of Jewish agricultural colonies occurred from 1881 through about 1940, when during that time period over 100 colonies were created.

The earliest of these agricultural experiments were started by German Jews, but with the onset of Russian pogroms an effort to relocate, fund, and settle Russian Jewish immigrants in agricultural colonies in the United States began in earnest. This movement coincided with an increasing urbanization of North America, creating access to an increasingly depopulated and abundant farming acreage often in remote locations spread throughout the country. As Moses Rischin explains of these demographic shifts: "Our understanding of the first era of Jewish continentalism would be incomplete . . . without some appreciation of the most dramatic, inspiring, and disastrous venture in the whole history of the Jewish westward migration: the founding of agricultural colonies" (1991: 37).

Three basic problems led to the downfall of almost every attempt at creating these Jewish farming communities. These were "lack of farming expertise, poor soil, and insufficient funds" (Bartelt 1997: 354). Despite these failures, as Robert Goldberg points out,

> Few American Jews realize . . . that they need not look back thousands of years to their biblical ancestors nor across thousands of miles to the Israeli *kibbutz* to come face to face with their *Jewish agrarian heritage*, for no group in modern America was more obsessed with the agrarian idea in their fashion than their forefathers who between 1881 and 1915 founded over forty agricultural colonies across the length and breadth of the American continent. This collective effort in America was ideologically inseparable from an international Jewish Back to the Soil Movement that saw Jews establish dozens of farming settlements that extended from Argentina to Palestine, from Russia to Canada as well as in America and elsewhere. (1991: 69–70; emphasis added).

Goldberg continues, positing that "despite [their] failure[s], the utopian dreams that informed many of the pioneers who invested their lives in [Jewish] agricultural colonies are a tribute to the social idealism that has illumined many phases of American Jewish history and that has made its most pronounced contribution to the labor and socialist movements" (ibid.: 68). This social idealism is now interacting with and contributing to environmental and modern religious agrarian movements and is built on the taproot of Sinai consciousness and historic support for progressive, liberal politics that largely defines Jews in the United States.

This historic legacy of liberal Jewish politics and active engagement with social movements has created the ferment for the development of the robust, emergent Jewish religious agrarianism elucidated in this book. Besides this political and agricultural background, there exists a sophisticated and bourgeoning Jewish (especially US) approach to environmental issues. The environmental ethics of this movement have equally helped lay the groundwork for the Jewish agrarianism of Hazon and the various synagogues and temples that join CSA partnerships under their aegis.

JEWISH ENVIRONMENTAL THOUGHT

Judaism includes the potential for a strong anti-environmentalist strand of thinking. This is exemplified especially in Orthodox circles, where *midrashim* and *halakhic* reasoning tend to privilege a male-centered legal tradition that sees Yahweh's creation not as a biocentric place of reverence but where an ongoing divine covenant with humans was put into place. Within this legalistic and religious perspective, human action and fidelity to God's laws is paramount, and concern for nature can be interpreted as idolatry. Nonetheless, "Biblical law specified in great detail the proper treatment of the soil, animals, and vegetation of the land of Israel in order to maintain Israel's religious ritual purity and moral integrity" (Tirosh-Samuelson 2006: 27).

With the destruction of the Second Temple in 70 CE and the development of rabbinic Judaism, Torah became the center of Jewish life, and its concern with ritual purity and moral integrity moved from the inner sanctum of the temple to how Jews were to live in diaspora. This privileging of Torah meant that "rabbinic Jews would experience the natural world through the prism of Torah" (Tirosh-Samuelson 2006: 28). The twofold pressures of both living in secluded ghettos away from farm land and the fear of idolatry contributed to a prism that viewed the natural world to be, at best, a peripheral concern for most rabbis.

This understanding of rabbinic Jewish history, coupled with the history of displacement from the land discussed earlier, leads to a second starting point in regard to contemporary Jewish environmental thought. Because rabbinic Judaism has its focus (religious, legal, intellectual, social) on Torah, the inclination to "mine" texts so as to apologize for Jewish complicity in the ecological crisis and to provide evidence of environmental concerns has been tempting. This approach leads to what Bradley Artson describes as a "contextless collage of Rabbinic and Biblical sayings" that avoids "the more comprehensive consideration of how Jewish tradition might respond to the larger question of living responsibly with nature" in today's world (2001: 162). In this Judaism is not different from other text-based traditions, including varieties of Christianities that privilege the Bible. Rather than a decontextualized, apologetic, externally pressured to seek for green insights, and finally meaningless mining, Artson argues that Jews need to create a second stage of contemporary Jewish environmental thought,

> one which explores helpful conceptions of the Earth and humanity from within the context established by Jewish thought and writings. Although our tradition may say little directly about air pollution or about the polar ice caps, it does dwell at great length on how Jews are to live with the soil, on the sanctity of the Earth and its produce, the holiness of one particular place (*Eretz Yisrael*), and on particular times (*Shabbat, Yovel,* and the *Shemittah*). Perhaps if we begin with intrinsically Jewish categories we might construct a Jewish ecology, enriching the traditional structure of Judaism with a consciousness of environmental issues rather than simply tailoring Jewish religion to fit within the procrustean bed of a dismembered ecological Judaism. (ibid.: 162)

Thus, a vibrant, articulate, contemporary Jewish environmental ethic must first speak to a unique culture and history rather than plunder them to find token passages and legal examples that can be applied to larger environmental concerns. According to Artson, a second stage will create a true, possibly systematic environmental ethic that is firmly grounded within Jewish cultures, cosmologies, and ontologies. His redaction hints at some of the key agrarian concerns shared with me by members of Hazon and Koinonia: concerns for proper living relations with the soil, the sanctity of Earth and its caloric bounty, and the centrality of place.

A central aspect of this Jewish cosmology is Israel—the land promised to Jews from God. Because of the diaspora, written and oral Torah are also central to Jewish cosmologies. This means that while God commands various *mitzvoth*, Jews are also obligated to say blessings (*berakhot*) of gratitude for the various gifts God has promised and shared with His creation. Such historical and material gifts begin with the literal land of Israel and its bounty as described in the Hebrew Bible. The Torah, Artson explains, "prescribes gratitude to God as Sovereign of Israel; the Rabbis extend that response to the world" (2001: 167).

From this rabbinic perspective, the Earth is the Lord's, and by saying blessings the various created objects of the Earth encountered outside of the Holy Land of Israel are made sacred and, consistent with the original covenant of Israel as the Promised Land, are thus given over to Jews to use. This legal tradition of saying blessings "lent significance to the soil, to what emerges from the ground, and to the sanctity of all ground everywhere. The categories of the *berakhot* themselves valorize the Earth as a sacred and sustaining presence. . . . They also enforce the insight that a sacred place is anywhere that life can thrive, that a sacred time is anytime Jews gather to eat and pray" (Artson 2001: 167–68).

Artson's historical analysis of Jewish blessings and how this process widened the circle of which landbases could be counted as holy helps create an internal Jewish ecological ethic, one that can be argued is explicitly agrarian. We have seen that one of the guiding goals of environmental agrarianism is precisely to create sustainable farms where a diversity of life can thrive. According to Artson's interpretation of the rabbinic tradition of reciting various Jewish blessings, these farms (especially if farmed by Jews) become hallowed ground if and when properly blessed. Such sacredness is compounded when the food grown on the farms is blessed, and compounded further when the blessed food is consumed with and prayed over by other Jews.

If we build on Artson's argument for a second-stage Jewish environmental ethic/environmental thought, then we are able to return to the political tradition alluded to earlier. This is a tradition that compels Jews to strive for justice. As Richard Schwartz explains, "Jews are required to protest against injustice and to try to agitate for change even when successful implementation appears very difficult" (2002: 3). Given the various issues of justice in regards to food production—in other words, taking seriously environmental agrarian critiques about how farm workers are treated, how farm animals are treated, how the soil is treated, and how farmers are treated in the corporate-controlled food system—Jews are required to take food justice issues seriously. The epigraph for this chapter details one religious agrarian rabbi's belief that more and more Jews will precisely take environmental justice issues seriously in the coming decades.

Of course, this is an ideal position for which Schwartz argues, built on the lineage of Shabbat 54b found in the Babylonian Talmud. This passage states, "Whoever is able to protest against the transgressions of his own family and does not do so is punished [liable, held responsible] for the transgressions of his family. Whoever is able to protest against the transgressions of the people of his community and does not do so is punished for the transgressions of his community. Whoever is able to protest against the transgressions of the entire world and does not do so is punished for the transgressions of the entire world" (Schwartz 2002: 1). Given the inherent complexities of scale, complicity, and interconnectedness of environmental justice issues, it is impossible for any one person, regardless of religious identity, to always act in a just way. However, using food as a locus of putting into practice a Jewish environmental ethic allows for an attempt to be made. The Jewish concern for justice indeed motivates many people who have joined Hazon-sponsored CSA arrangements.

Let me briefly return to Schwartz's book because it is a sustained attempt to develop Jewish environmental ethics given the ecological crises facing the planet and is important to my overall argument. According to Schwartz, Judaism contains within it strong biblical, prophetic, and legal statements that require adherents to be actively involved in justice. Jews should be concerned with ethical issues that range from loving one's neighbor (Leviticus 19:18) and Hillel's version of the Golden Rule (what Schwartz calls in Hebrew *gemilut chasadim*, or "loving kindness"), to social justice and the need to both act righteously (tzedakah) and give charity, to concerns about peace, international relations, hunger, and ecological issues.

Regarding environmental justice issues, Schwartz argues, "Many fundamental Torah principles express and make concrete the Biblical statement, 'The earth is the Lord's and the fullness thereof' (Psalms 24:1)" (2002: 39). According to him, this statement means that people (especially Jews) are to be co-workers with God in helping preserve His creation since everything belongs to God. Another passage that Schwartz alludes to, and many other leading Jewish environmental thinkers use as well, is Deuteronomy 20:19–20, which contains the prohibition to not unnecessarily waste or destroy (*bal taschit/tashchit*) anything of value, including especially fruit trees.[9] With this *midrash* of Torah we see within the rabbinic tradition a bourgeoning development of a concern for place, and this concern is shared by Jewish religious agrarians, as seen in chapters 4, 5, and 6.

The emerging analyses of Jewish tradition that support these concerns are happening because activist Jews argue that corporations have commodified the planet and have lobbied governments to craft free trade agreements that buttress a global consumer economy, such that the planet is in peril. Schwartz's prescription for this diagnosis is one "based on biblical values of 'The earth is the Lord's' and *bal tashchit*. [Jews] must work to change the current system, which is based primarily on greed and maximization of profits and entices people to amass excessive material goods, thus causing great ecological damage. We must work for approaches that put primary emphasis on protection of our vital ecosystems" (2002: 52–53). This prescription and analysis of the current ecocrisis is consistent with those put forth by environmental agrarians and advocates of sustainable farming practices (as well as New Monastics). It is an analysis contextualized, however, within a biblical and especially Jewish context. It should be noted that such exegesis of Jewish scripture is proliferating and is helping marry religious and environmental agrarian concerns within the lived practices of Jews in the

United States. We encounter these calls for justice and other explicit Jewish environmental ethics in later chapters because these are driving concerns for many US religious agrarians.

A US JEWISH VISION OF TRANSFORMATION

Jewish environmentalism today is generating a more visible presence, as reflected in the formation, development, and growing presence of Hazon, which has quickly become the leading voice for Jewish environmental concerns and the largest Jewish environmental group in the United States in terms of membership. Beginning as a small nonprofit in the Northeast in 2000, Hazon now has its main offices in New York City and affiliate offices in San Francisco, San Diego, Philadelphia, Denver, and Boulder, Colorado, and has merged with the Isabella Freedman Jewish Retreat Center (see end of this chapter for more on this merger). The creation of founder Nigel Savage,[10] Hazon currently has a board of directors that includes various rabbis and doctors, an advisory board, chairs of various campaigns, rabbinical scholars, and a staff of full-time employees who are supported by various program fellows. Hazon's catchphrase signals the group's anchor in Judaism and reads, "The Torah is a commentary on the world, and the world is a commentary on the Torah."

Hazon hopes to change, renew, and guide Judaism in the United States (and Israel) by focusing on creating transformative experiences, providing leadership in Jewish education and philosophy, and contributing to capacity building by providing grants, organizational support, and joining Jewish networks.[11] The two main campaigns that define Hazon's work are bike rides (mainly held in New York, San Francisco, and Israel) to raise awareness about environmental issues and contribute to the emotional, physical, and environmental health of Jews, and work on helping US Jews and the larger national society develop robust, sustainable food habits and networks of sustainable agriculture. Both of these issues are of central import to Savage, who is an avid biker and supports the ideals and goals of sustainable agriculture.

To this end Hazon has published a sourcebook that outlines and defines their vision of sustainable food systems, called *Food for Thought: Hazon's Sourcebook on Jews, Food and Contemporary Life* (Savage and Stevenson 2009). The book highlights Hazon's work on sustainable food issues, with eight chapters: Learning Torah; Gratitude, Mindfulness and Blessing our Food; Kashrut; Bread and Civilization; Eating Together; Health, Bodies and Nourishment; Food and Place; and Food and Ethics. Many chapters address explicit agrarian concerns and represent concerns shared at the level of religion-in-action in the lived networks of Jewish religious agrarians who are involved with Hazon-sponsored CSA partnerships. These are also concerns shared by members of Koinonia, but for the Jews I research, the concerns are germane to a unique Jewish tradition (for more on this, see chapters 4, 5, 6, and the conclusion). The overlap that definitely exists in Jewish and Christian religious agrarianism in the United States occurs as a result of environmental agrarian concerns as well as from the growing ecological reformation found within many pockets of the world's "dominant" religions, including Judaism and Christianity.

A fellow of Hazon's Food Program, Natasha Aronson, included a cover letter with my copy of *Food for Thought* that asserts:

> Rooted in more than 3,000 years of Jewish tradition about food, Hazon has inspired a growing conversation in the Jewish community about what we eat, where our food comes from, and how our food choices are so closely linked to our personal health and the health of our planet. Hazon is using food as a platform for Jewish education, drawing on the wisdom of our Jewish texts and sages for guidance as we confront contemporary food and environmental issues. Our food programs are designed to fulfill the Hazon mission: creating a healthier and more sustainable Jewish community, and a healthier and more sustainable world for all. People of all ages, from across the spectrum of Jewish practice and denominations—as well as those who are unaffiliated with the Jewish community and those from other faiths—participate in our food programs. (emphasis added)

This communication strategically confers legitimacy to Hazon's work by placing it within the larger tradition of Judaism. It explicitly mentions the "contemporary food and environmental issues" that define environmental agrarianism, including its religious aspects. This agrarianism, strongly associated with Wendell Berry, heavily criticizes the global industrial food economy and attendant fertilizer and chemical input monoculture farming practices that critics claim are destroying farming communities and soils. Furthermore, the passage articulates the mission that has guided Hazon since it was formed: creating healthier and more sustainable Jewish communities and a healthier and more sustainable global community.

It is important to realize that Hazon is not isolated from other Jewish organizations. Rather, its members and leaders are active in a variety of communities and form networks with other religious communities, environmental NGOs, politicians, and governing bodies. Hazon identifies as a progressive, articulate, environmental group, one that is thoroughly grounded in Judaism. Members include Orthodox, liberal Reform, Conservative, Renewal, Reconstruction, Zionist, and cultural Jews, and Hazon is even welcoming of atheists and agnostics (though the latter tend to be in the extreme minority of members and typically have a Jewish background).

Such inclusivity of Jews from various backgrounds led Hazon founder Nigel Savage to assert that Hazon "works to create a healthier and more sustainable Jewish community—as a step toward a healthier and more sustainable world for all" (2007: xiii). To further the creation of more sustainable food practices, Hazon organizes CSAs so local sustainable farmers can find customers and partner with local Jews (and non-Jews, for it is not necessary to be Jewish to join a Hazon-sponsored CSA). They also organize and host an annual Food Conference (which I attended in December 2009 and December 2013) dedicated to exploring sustainable food issues from a Jewish perspective. Hazon hosts a food blog called *The Jew and the Carrot* (http://www.jcarrot.org), which contains a clearinghouse of reflections about and analyses of sustainable food issues, all written by Jewish bloggers. Last, as the photocopy I received from Aronson describes, Hazon has created the "Jewish Food Education Network [,] a network of Jewish food education resources, curricula and teacher training including *Min Ha'Aretz*, an 18-lesson interdisciplinary curriculum about food and Jewish tradition for students (typically grades 5–9), a complementary curriculum for families, and additional training and resources on *Food for Thought*."

Sustainable food issues have indeed become Hazon's main concern. Efforts to advance this concern include the education of young Jewish children by teaming up with Jewish Day Schools, the development of CSA partnerships, generating a sustained effort to reinstitute the *shmita* year every seven years with its implications for agriculture,[12] and the creation of a vibrant online community at *The Jew and the Carrot*. Altogether Hazon is directly speaking to and meeting the growing demand and needs of US Jews who are concerned about sustainable food issues. Their efforts situate them as the leading voice for sustainable Jewish food issues in North America.[13]

This concern and call is shared in the preface to *Food for Thought*, which states in agrarian yet equally Jewish terms that

> How we eat is an extension of who we are. How we consume is a central manifestation of our relationship to the world. . . . The conversation about how we eat in a way that is enjoyable, sustainable

> and ethical; in a way that engages our families and our tradition, and also the unique circumstances of the world that we live in—that conversation, in our view, enriches Jewish life and makes the world a better place. (Savage and Stevenson 2009: xi)

Hazon takes seriously the starting premise that inspired me to write this book: eating matters. Furthermore, the choices of what we eat, how we eat it, and how the food that we eat is grown all assume religious significance for religious agrarians, including those exemplars in Hazon. Food choices that exhibit concerns about justice, ethics, and sustainability are being grafted onto existing Jewish traditions and lifeways so that Jewish identity is enriched. Moreover, this bottom-up reimagination of tradition captures the essence of lived religion in action. It places Jewish religious agrarians in the trophic pyramid of calories. Hazon's Jewish dietary choices reflect cosmological views about values and practices in regard to the natural world. They are also centrally concerned with many of the goals of sustainable agriculture. By taking sustainable agricultural issues seriously and merging these with Judaism, Jewish religious agrarians in Hazon recognize that they are part of a more-than-human world where agricultural production is seen to have either positive or deleterious impacts.

The preface continues, stating

> And the interplay between the two elements [agrarian and religious] of that conversation is one that we want especially to encourage . . . we share a love of food and of learning, and the deep belief that learning about food and doing so through the double prism of Jewish tradition and contemporary challenges is vital to the creation of healthier relationships to food in the broadest sense. . . . Our dream for the whole world? Abundant and healthy food for everyone, food that isn't produced at the expense of people or the earth, food that builds communities and encourages us to be our best selves. (Savage and Stevenson: xi)

The sustainable farming practices supported by Hazon participants are largely inspired by their desire to be their best Jewish selves and active participants in building local food culture and communities that nourish farmers, soils, and bodies. By entering into a living conversation about these issues, Hazon is helping change the politics, religious identities, and food habits of the Jewish foodscape in the United States.

SHEARITH ISRAEL

Shearith Israel is an egalitarian Conservative synagogue that occupies a brick building built in 1904 located near the Briarcliff Campus of Emory University in Atlanta, a part of the city that is mostly affluent and Caucasian.[14] Located down a tree-lined side street and surrounded by larger family homes, the synagogue recently renovated its building, installing eco-friendly light bulbs throughout their campus and flush-free urinals in the men's bathroom. Part of the campus is rented out two days a week to a Jewish children's school, to whom they serve kosher meals made in a kosher kitchen maintained on site. Within three miles of the synagogue are a Whole Foods natural food supermarket; Sevananda, a member-owned natural foods co-op; a kosher gourmet-themed deli and store; Judaica Corner, a retail store that sells material artifacts (books, Talmuds, menorahs, clothing) of Jewish religion and culture; and the retail store for Daron "Farmer D" Joffe, a leading proponent of Jewish religious agrarianism and member of Hazon who sells sustainable farming tools and composts. Also within a short driving distance is Little Five Points, a trendy part of Atlanta that is home to cafes, restaurants, and an independently owned bookstore.

On one level, this is a far cry from the rural Deep South farming communities of Americus and its surrounding environs that is home to Koinonia. However, industrialized farming has its critics in areas as diverse as Americus and Atlanta. This is because the criticisms of industrialized farming developed by Berry and others apply to both rural and urban areas, for in both areas the values and practices of food choices of the average citizen predominantly support industrial farming.

Furthermore, the sustainable agriculture worldview is supported and developed by scholars, activists, farmers, and researchers who live in places as diverse as Salinas, Kansas (home of the Land Institute), to Atlanta (home of Farmer D), to Will Allen of Growing Power based in Milwaukee and Chicago. Due to the "unsettling of America," to use Berry's book title (1977), where less than 2 percent of citizens are now involved in farming (compared to about 75 percent at the time of the Civil War), the majority of people concerned about sustainable food issues are now by default urban dwellers. Such demographics help make sense of the growing popularity of farmers' markets and CSAs in urban areas and the development of religious agrarian concerns in urban religious communities.

During my first visit to the synagogue, at which time I interviewed the lead rabbi who was supportive of hosting a Hazon CSA partnership and during which I observed a Wednesday afternoon Tuv Ha'Aretz pickup, I was able to meet with a past president of the synagogue. My discussion with this member ranged from discussing the history of the congregation to its contemporary identity to food and environmental issues. According to Mike Greenberg, an articulate, soft-spoken man of late middle age who has been a member of the congregation for thirty-one years, the synagogue actually began as an Orthodox community in 1904 under the leadership of Rabbi Tobias Geffen. In 1957 it switched to mixed seating, and about thirty years ago it became "traditional," in the sense that the members were ritually observant but the rabbi decided the level of this observance. In 2001 the community became Conservative and egalitarian, in that women can read Torah (they have a female rabbi) and women can be included in the *minyan* needed for quorum.

Greenberg also explained that the synagogue is eclectic in its makeup, ranging from an eighty-five-year-old Holocaust survivor to professionals with young children. He feels there is no division within the synagogue regarding the CSA and Rabbi Joshua Hillel's vegetarian views (discussed in later chapters) and that the CSA group and rabbi tend to be more progressive in their Jewish and political views than the rest of the congregation. Greenberg was involved with the hiring of Rabbi Hillel, claiming he was hired for his humanist teachings and that environmental issues were not talked about in the hiring process. Nonetheless, since the rabbi began in 2006, environmental concerns have been brought to the attention of the synagogue in large part because of his progressive environmental views and support of hosting a Hazon-sponsored CSA partnership.

Greenberg claims that of the 550 member families in the congregation, most have been practicing Jews their whole lives or are from a Jewish background. There are a few converts, most from Christianity, while some converted from Orthodox Judaism because of the perceived exclusivity toward women in Orthodox teachings and practice. The synagogue is not liturgically liberal, as approximately 95 percent of the services are in Hebrew. Furthermore, more congregants have a desire to deepen their understanding of Judaism and Jewish traditions so a second female rabbi was hired to help hold Torah, *mishnah*, and Talmud classes. Greenberg explained that the synagogue as a whole is supportive of the CSA arrangement, but the majority of the synagogue are not Tuv Ha'Aretz members.

My first Wednesday visit to the synagogue coincided with one of the final Tuv Ha'Aretz CSA drops of the summer season at which there were thirty-two boxes present. Of these CSA members, approximately twenty-five belong to the synagogue proper; the rest are Jews who are not full-time members or are friends of CSA members and are not Jewish. The organizer of the CSA partnership who schedules the Tuv Ha'Aretz arrangement and Wednesday pickup and activities is Jewish and an active member of Hazon but not a member of Congregation Shearith Israel (CSI). She approached Rabbi Hillel about using the synagogue as a place to start a Hazon-sponsored CSA, to which the rabbi agreed, largely because he is friends with Hazon staff in New York (where he and his family lived and worked before relocating to Atlanta) and because he is an active voice in Jewish vegetarian movements. This arrangement is further explored in the subsequent chapters as I use data gathered from interviews with the rabbi and CSA organizer to explore Hazon's religious agrarianism.

LIVED JEWISH FARMING NETWORKS

Consistent with my argument about seeing emerging religious agrarian and larger religious environmental concerns as giving evidence of a lived networking within religious groups, Hazon maintains numerous active network partnerships with other North American Jewish environmental groups, NGOs, and nonprofits. These include supplying Hazon-sponsored grants to emerging groups that are working on environmental and especially sustainable farming issues. One such group that has received a grant and that shares web links with Hazon is the Jewish Farm School, which works toward "sustainable agriculture rooted in Jewish tradition."[15] Begun in 2006 with help from Hazon, the Jewish Farm School runs summer camps and internship programs for young Jews in high school and college to learn about sustainable farming on successful organic farms. As their website states, "Our emphasis is on teaching practical skills while also educating about the larger context of our contemporary food systems, and how our traditional values and practices can inform our decisions and actions today."

Here we see religious agrarian networks in action that explicitly use religious teachings and values to actively shape and change the values and practices of religious individuals in regard to sustainable food choices by providing hands-on training and education regarding environmental agrarian concerns. These lived networks also provide a US context within which Jews are reinterpreting and renewing their tradition to address contemporary sustainable food issues. Such a pivoting of the sacred is a sign of religious agrarian work in action, with Hazon and their rhizomic lived network relationships helping lead this movement. Hazon maintains further active networks with other emerging progressive Jewish environmental groups and activity centers. These include Wilderness Torah (http://www.wildernesstorah.org), Eden Village Camp (http://edenvillagecamp.org), and the Jewish Climate Change Campaign (http://www.jewishclimatecampaign.org).

Other networks with which Hazon cross-fertilizes include the Teva Learning Center, which bills itself as "North America's foremost Jewish Environmental Education Institute" (http://tevalearningcenter.org/). Teva has become one of the leading groups conducting on-site environmental education for Jews, and many of their staff and former interns are now active with Hazon. The same holds true for former interns and staff of Adamah, an organic farm located near the Isabella Freedman Jewish Retreat Center in Connecticut.[16] Adamah runs three-month summer interns for young Jewish adults who want to learn more about organic farming. As their website states, "Programs at Adamah integrate physical, social, spiritual, Jewish and ecological realms in order to inspire participants to a life of service—to the Jewish community and to the earth." The Isabella Freedman Jewish Retreat Center hosted Hazon's 2010 Food Conference East and the 2013 annual food conference I attended.

Moreover, and tellingly, Isabella Freedman and Hazon merged into one organization in 2012. This key event was shared by Nigel Savage, who emailed his thoughts about the merger to the Hazon online discussion list on December 3, 2012. He writes,

> Sometimes our lives cross and crisscross in unexpected ways. Things that one thinks are incredibly significant prove minor; and minor or accidental decisions can change one's life. So it was for me in the summer of 1998. That was when, for the first time, I visited Isabella Freedman; spent time at Elat Chayim (then in Accord, NY); and met people like Adam Berman (who went on to direct Isabella Freedman and to found Adamah), Reb Zalman Schachter-Shalomi, Nili Simhai (currently the director of Teva) and Rabbi Dr Arthur Waskow [one of the elders in North American Jewish environmental circles]. Not more than a few moments' thought went into the serendipitous construction of that summer. But encountering those people and institutions led me to found Hazon, late the following year. And today, things go full circle: we're announcing that Hazon is merging with Isabella Freedman (which includes Elat Chayyim, following a merger a few years ago). The Teva Learning Alliance, until now a part of Surprise Lake Camp, is also going to be part of the new Hazon.

When providing the reasons behind this merger with Isabella Freedman, Savage writes,

> We're both significantly engaged with leadership development. How do we re-frame what Jewishness is or could be? How do we tease out the places where Jewish tradition's understanding of how we relate to the natural world critiques some of what today we assume is "normal?" How do we connect and network young leaders so that they can support and encourage each other for a lifetime to come? This area includes Hazon's work in curriculum development, the support we provide to the growth of Jewish CSAs, the platform we've created for the Jewish Food Movement overall, and the determined work we're already doing in preparing for the next Shmita year in 5775 (September 2014). For Freedman this includes their Jewish Greening Fellowship, spurring institutional transformation through leadership development; and it especially includes the Adamah and Teva programs, whose alumni fuel vibrant Jewish life of all sorts across the country.

Notice that Savage directly mentions that the strategic joining under the larger Hazon rubric of Judaism's leading environmental groups in the United States (and North America, broadly) is done largely to further their religious agrarian concerns and goals to continue growing the "Jewish food movement." He also speaks about the need to empower and educate the next generation of North American Jews, young people who are eager to learn more about the planet and what they can do to live sustainably on it. This is a recent concern given Jewish history and traces itself back to the onset of the larger US and global cultivation of an ecologic consciousness. Yet this concern is grafted on to religious identity. Savage speaks to my argument in chapter 1 about religious values and political activism, especially for Jews in North America. This is evident when he points out that

> Overall: I'm incredibly proud of what together we've accomplished in the 13 years since the Nash family first gave me the money to found Hazon. But I also note, overall, the challenge of asset allocation in American Jewish life. The installed capital base of the American Jewish community—all the synagogues, day schools, camps—is probably worth more than $10 billion. The total value of family

> foundations established by Jewish families is almost certainly in excess of $50 billion. Yet the total revenues of all the Jewish green organizations—broadly defined—is barely $10 million. That's probably up 10-fold in a dozen years, but it's still less than a fraction of one percent of the annual revenues of American Jewish life. Given the vitality of food, the environment and the outdoors as profound gateways to Jewish life and to social change, that just doesn't make sense—and we'd like to help change that in the coming years.

Here we see that the vision of Hazon, along with its strategic merger and the active lived networks it has developed and supports is to continue being the leading voice of environmental values for North American Jews. To undertake this work will require infrastructure, including financial support—hence Savage's appeal. Even more, when explicitly justifying this merger, Savage tellingly writes:

> So if many things will not change because of the merger, why are we doing it? The simplest answer is: *More impact.* In the last two or three years it has become clear that after 15 years of investment in "start-ups" in the Jewish community, we now need to devote different sorts of resources to enable a smaller number of organizations to really go to scale. Doing this is vital if we're serious about transforming American Jewish life (let alone enabling the Jewish community, as a community, to have any impact on some of the greatest issues of our time, such as developing sustainable food systems, or seeking to prevent even worse climate-related destruction for generations to come). (emphasis in original)

This text speaks to the growing strength of the Jewish food/environmental movement and the weakness of having to bring about structural change in the lives of Jews and non-Jews in a larger individualistic society that is not focused on sustainability in agriculture and lifestyle. How to remedy this? For Savage and others in leading North American Jewish environmental groups, it is to make more secure lived networks with goals of further education and capacity building to deal with key issues like climate change and cultivating sustainable food systems. Yet this will take money and political lobbying, a struggle that also faces Koinonia and other minority voices of US religions. These voices have a strong concern about sustainable living but are historically the minority perspective within religious traditions and often struggle to make ends meet.

Despite this challenge of maintaining infrastructure, Savage recognized the amazing growth in the past decade and concludes, sharing, "This is important because the growth of the Jewish Food Movement, of Jewish outdoor education, and of Jewish environmental education in the last 10 years is incredibly profound—and it's still incredibly under-resourced. Under the radar of American Jewish life,—and despite a community that is in overall terms trending flat to trending downwards, by a variety of metrics—remarkable things are happening." Notice his excitement about real changes he has encountered and witnessed. Notice also the timeline—the previous ten years, which supports my thesis that religious agrarianism is a recent but rapidly growing element of religious environmentalism in North America. Savage ends by saying,

> We're weaving the Jewish community of tomorrow, and it's happening today. The kids who go to camp at Eden Village or Ramah Outdoors, who go with their school to Teva, who go on to do a trip with Jewish Farm School, who teach at Teva, who live at Adamah or Urban Adamah—these are people who take for granted that there is no distinction—and should be no distinction—between learning and growing Jewishly, between celebrating our ancient tradition, and between being good citizens of our small fragile planet.

Savage's excited announcement predicts the same future that Rabbi Hillel predicts: more engaged Jews taking sustainable food issues seriously in the years and decades to come.

The common concerns and active networks maintained by these groups, with Hazon strategically situated in the middle, are changing the values and practices of environmentally concerned Jews in North America. In the chapters that follow we meet some of these people and hear their own words about environmental and religious agrarian issues. I explore the concrete actions they are undertaking in the biological landscapes of sustainable agriculture fields, acting directly or through joining a CSA.

For example, the 2010 Food Conference East had workshops on permaculture, "Organic Torah: How to Think Like an Organic Farm," "Jewish Values, Fair Trade, and the Highest Level of *Tzedekah*," and a hands-on class on goat milking and cheese making with farmers from Adamah. The 2013 Food Conference at Isabella Freedman had presentations and workshops on topics ranging from "Shmita: What it Meant Then, What is (Could) Mean Today" to

"Brewing Up the New Year" to "Peeling Back the Celophane Veil: The Kosher Meat Industry," and with clear connections to religious agrarian concerns for justice, presentations on "Farm to Fork: A Look at Struggles throughout the Food System" to "Lay of the Land: Exploring Jewish Food Justice Programs," to "Now What? Exploring the Strategic Jewish Role in the Food Justice Movement." Last, like Koinonia, Adamah Farm is also practicing permaculture, which I learned about by participating in a tour led by Sarah Chandler, the farm manager, and Yigal Deutscher, a leading Jewish permaculture practitioner who is leading Hazon's Shmita campaign.

Such diversity of topics and concern can possibly lead to burnout and/or diffusion of focus among Hazon members. This has led Mark Jacobs to recognize that there are external and internal challenges to the development of a robust Jewish environmentalism, especially because "A diversity of religious cultures and a wide range of levels of religious and cultural knowledge create a challenge for organizing meetings and conferences at which all participants feel equally comfortable and empowered. Fortunately, most of the people who have been attracted to Jewish environmentalism have been eager to work through these challenges" (2002: 473). My experience at two food conferences is that Hazon makes a very concerted attempt to overcome possible sources of conflict within a small (growing) active North American Jewish environmental community. The active networks maintained by progressive religious environmentalists provide a mechanism to work on overcoming possible sources of theological discord and create an inclusive movement. This is evidenced by a quick glance at the literature shared at the 2013 Food Conference, where I saw publicly displayed and readily available pamphlets, magazines, flyers, and handouts from Grow and Behold, a kosher pasture-raised meat company started by Hazon affiliates; Jewecology.com; Organic Torah; *Joy of Kosher* magazine; Elat Chayyim, the Center for Jewish Spirituality; Green Zionist Alliance; and other handouts provided by Hazon or conference participants, including handouts on LGBTQ lifestyles, and a variety of worship services held each conference day. This broad acceptance is found on Hazon's literature provided at the conference, where on a half-page handout, they write,

> How do you apply 3000 years of Jewish wisdom about food to contemporary food conversations? As Jews, we've been thinking about kashrut—about what is "fit" to eat—for nearly 3,000 years. A growing number of people today realize that our food choices have significant ramifications—for ourselves, our families and the world around us. Hazon uses food as a platform to create innovative Jewish

> educational programs to touch people's lives directly, to strengthen Jewish institutions, and to create healthier and more sustainable communities in the Jewish world and beyond.

Hazon contextualizes its food advocacy within the larger Jewish tradition, and does so in a way that merges it with contemporary environmental agrarian concerns and in language that is inclusive of all Jews and non-Jews.

Jacobs continues,

> I have observed a broad array of ideas that lead American Jews to organize within their religio-ethnic community on environmental issues. They can be grouped into five categories: fulfilling Jewish obligation; fulfilling universal obligation; effecting broad cultural/political change; strengthening the Jewish community; and personal fulfillment . . . With rare exception, particular theological conceptions, halakhic (legal) obligations, and ethical analyses do not seem to be driving forces for the organizing of specific activities or programs . . . Most individuals involved in Jewish environmental efforts are motivated by more than one of these beliefs. (2002: 464)

We see that a mix of these do provide motivation for some to become involved with religious agrarian issues; but also, and significantly, ethical reasons are a motivating factor in Jewish religious agrarianism (see especially chapter 6 on justice).

This motivation was seen at the 2013 Food Conference, where on a Monday night there was a panel discussion, "The State of the New Jewish Food Movement." Deep into the discussion, Rabbi Noah Farkas, founder of Netiya, a Jewish interfaith partnership working on urban agriculture and food justice issues in the Los Angeles region, exclaimed that "The [Jewish] food movement has to have a bar mitzvah. It has to grow up. There has to be an engagement with power." Farkas also quoted Rabbi Abraham Joshua Heschel, who stated, "Few are guilty, all are responsible."

So what does a grown-up, responsible, Jewish and Christian religious agrarianism resemble in the twenty-first century when so much is at stake? To answer this I turn to an exploration of religious values and their influence on religious agrarian practice, focusing on one of the three emergent religious agrarian themes my research discovered: concern for "the local."

CHAPTER FOUR

THE LOCAL ([FARM] LAND)

> The agrarian standard, inescapably, is local adaptation, which requires bringing local nature, local people, local economy, and local culture into a practical and enduring harmony.
>
> —Wendell Berry (2003: 33)

> Locally Grown: Thousands of Miles Fresher.
>
> —Popular bumper sticker

> We take that very real image of the farmers in the field, and of course I make it into a metaphor of counting the bounty in our lives and being aware of what is in our lives. And even though our congregants are doctors and lawyers and professors and dentists, we can still have that imagery of the land very present in our own lives and use it as metaphors to enrich our lives even though we may not be farmers ourselves.
>
> Q: And do you get affirmation from [congregants]?
>
> A: Sure, absolutely.
>
> —Rabbi Greenberg

LOCALIZING AGRARIANISM

So far in these pages we've been exposed to answers about the questions: why religion, why values/ethics, why food, why sustainability, and why in the United States? We've also explored key concepts behind sustainable agriculture, critiques of industrial agriculture, conceptions of place, and the basics of agrarianism. Last, we've met my key research communities and the methodological and theoretical tools I used while generating the concept of religious agrarianism. In this and the next two chapters we gain insight to the three themes that mobilize the religion-in-action of religious agrarianism. The first key theme is concern for the local, the focus of this chapter. I also explore links between concerns for the local, how this concern manifests in sustainable agriculture and environmental agrarian worldviews, and how these concerns influence religious agrarian concepts of locality.

By analyzing research data from members of Hazon and Koinonia, we are able to see similarities and differences in the values behind how Jewish and Christian religious agrarians include concern for locality—local farmers, local farmlands, and local economies—in their lived religion. Such perspectives about locality have led Hazon to become the leading religiously inspired CSA network in the United States, as well as Judaism's strongest proponents of the religious agrarian concern for the local. Meanwhile, Koinonia is an exemplar of religious agrarian concerns for locality by definition: it is a community living at a local campus and practicing Christian religious agrarian concerns about locality.

Such concern for locality is present not just in religious agrarian worldviews but in contemporary North American culinary and food culture circles as well. For example, in the winter of 2011, the International Association of Culinary Professionals hosted their annual regional food conference. Part of it was held at the famed James Beard House and included a session titled "Farm to Table: Urban and Rural Foodways."[1] The website's description of this part of the conference reads:

> You care about where your food comes from. You have a relationship with farmers, fishermen and food purveyors who not only produce good food but [who] steward the environment. You believe that eating is a political act. In the last few years the local food movement has made enormous strides and gone from fringe to mainstream. What challenges do we still face in re-regionalizing production and shoring up our food systems? How can culinary professionals continue to bring good food to the table and expand access to healthy, fresh foods?

This description captures a rapidly developing environmental agrarian worldview and shows the increasing ubiquity of it as it percolates up into the leading edges of North America's culinary landscape. Among the most important issues endemic to this worldview are environmental agrarian themes of food politics; environmental stewardship; concern about knowing who grows your food, where, and how; concern for food health and quality; and "reregionalization," including at urban scales.[2] Important to notice is that central to these environmental agrarian concerns is the health of the land.

Environmental agrarians build on the larger agrarian tradition when they argue that without a healthy landbase from which to farm, it is not possible to have a healthy society, healthy body, or healthy body politic. I explore these themes in depth in the next chapter but since these concerns for health are built on a specific local place I begin with exploring the local.

Concern for the local is central to the work of leading environmental agrarian Wendell Berry, who writes,

> There are, I think, three questions that must be asked with respect to a human economy in any given place:
> 1) What is here?
> 2) What will nature permit us to do here?
> 3) What will nature help us to do here?
> . . . the first question has not been answered, or asked, so far as I know, in the whole history of the American economy. (1987: 146)

These questions reflect some of the key environmental agrarian ideas and ideals that constitute Berry's corpus.

Contemporary environmental agrarianism reflects this corpus, and overall environmental agrarianism includes concerns about fidelity to place, the desire to develop a learned husbandry that exhibits concern for locality, recognition of the agricultural and thus agrarian base of any and every human economy, and criticism of the industrial agricultural model of the United States that has never, as Berry charges, asked about the health of the land. Thus, knowledge of, concern for, protection of, and proper use of land are the paramount environmental agrarian concerns that should form the bedrock of a healthy culture. As Berry writes, the goal is to "regard the land as of ultimate value" (2005: 39). Such a sentiment spills over into the realm of theology as seen in theologian Paul Tillich's famous claim that religion deals with ultimate concerns and ultimate values. It also is directly religious, especially if we take religious studies scholar Ann Taves (2009) seriously when she argues that religion deals with special things set apart. Here "special things set apart" include the local and its relation with the sacred (both are special things and things set apart).

For environmental agrarians such concerns should ideally lead to the generation, over many years and decades (and eventually centuries), of a locally based farming culture that practices environmentally friendly and sustainable agriculture regimes (a further special thing set apart, especially from the globalized monocultures of industrial agri-culture). Such regimes will recognize that "local soil and local culture, are intimately related" (Berry 1990: 154) such that both have an economic and cultural value (ibid.: 157). Without a local culture based on a historical, intimate, and loving knowledge of "any given place," then "a place is open to exploitation, and ultimately destruction" (ibid.: 166). Environmental agrarians are concerned about the exploitation and destruction of local farms, local ecosystems, and local economies and recognize that these are interlinked with destruction at larger scales that is wrought by a globalized economy based on the use of fossil fuels. Many of the people I talked with at Koinonia and in Hazon share these concerns, so concern for "any given place" is a central aspect of religious agrarianism. We also see the reemergence of place in Western ontologies where the religious environmentalist processes of renewal, reinterpretation, and ecological hermeneutical retrieval are helping support such place-based focus on the local.

Religion and values are central to discussions about agriculture, place, and agrarianism. For example, Berry has written, "in talking about topsoil, it is hard to avoid the language of religion" (1990: 62). This statement is pertinent for my overall argument and, coupled with viewing land as the ultimate loci of values, allows for religion to factor into environmental agrarian concerns about the land. When Berry wrote this, he was discussing how topsoil retains and expels water; he was concurrently describing how topsoil is a living entity made from the dead, decaying bodies of countless organisms. For him this means that topsoil is not something humans can replicate. Rather, it is something made only by the "Great Economy" (ibid.) of nature and is needed for our survival. For environmental agrarians everywhere, we destroy healthy topsoil to our detriment. Conversely, if we build, maintain, and replenish healthy topsoil then we have the ability to flourish and build healthy cultures. The language of the sacred is necessary to discuss sustainable agriculture because society is literally the outgrowth of soil and land, even though we cannot fully comprehend how the miracle of soil works nor replicate this miracle in a laboratory.

Such religious language about soil is shared by the agrarian theologian and permaculture practitioner Fred Bahnson, director of the Food, Faith, and Religious Leadership Initiative at Wake Forest University School of Divinity. He writes that "soil is a portal to another world. . . .Working the soil opens us inward where we find a God eager to lavish upon us God's mercy and compassion and love. Soil also opens us outward, where we learn to receive the fruits of this good earth" (2013: 251). Philosopher and organic farmer Fred

Kirschenmann also eloquently writes about the sacredness of local soils when he states that, "in my theology, the divine always meets us in the flesh—*all* flesh—*all* relationships, not just our relationship with humans or relationships we like. This seems to me to be at the heart of the concept of the incarnation" (2010: 17). For religious agrarians, divine relation includes harmonious, gratitude-filled interactions with local soils on local farms.

Finally, Larry Rasmussen, the Reinhold Niebuhr Professor Emeritus of Social Ethics at Union Theological Seminary, shares a line from one of Henry David Thoreau's poems which reads, "Heaven is under our feet as well as over our heads" (2013: 196). In looking at ancient Hebrew society, Rasmussen interprets this to mean that "the soul is the animating element of our humanity and the way we touch the divine. But 'soul' can also be spelled s-o-l-e. Where is our soul/sole? On the bottom side of bare feet, in touch with the sacred beneath our sole, the soil. Soil and sole have always been in touch. So have soil and soul" (ibid.: 196–97). In these, and the sentiments and thoughts from religious agrarians shared below, we see that the local becomes the meeting place with the divine. For religious agrarians in the United States, this meeting place occurs in rapturous engagement with the divine, sacred soils of farm land, garden plots, and community gardens. As shared by Rabbi Noah Farkas at the 2013 Hazon Food Conference, "tending a garden is its own form of spirituality."

RELIGION, ETHICS, AND LAND

This section builds on the religious agrarian view of soil and presents a brief overview of key thinkers who have wrestled philosophically with what proper human–land relations might resemble, since by definition all our relations with the planet occur in a bounded landscape. When taken seriously, this means that by definition we are never not in relation with the land. The groundwork these leading thinkers helped lay became the ferment out of which Berry and other environmental agrarians grew. Their work also crosses over into the various ecotheological approaches to human–land relations that have developed in response to Lynn White's 1967 analysis of the current ecocrisis. With this history in place, it is possible to see how contemporary religious agrarianism is partly influenced by the history of this systematic exploration of human–land relations.

The first, sustained, articulate early ethical defense of land came from Aldo Leopold in his 1949 classic, *A Sand County Almanac*. In this collection of essays, the forester, ecologist, and restorationist opines, "It is inconceivable to me that an ethical relation to land can exist without love, respect, and admiration for land, and a high regard for its value" (1968: 223). Leopold maintains that our "Abrahamic" conceptions of the land, coupled with an instrumental economic system built on Lockean concepts of private property and labor value, have led to the despoliation of landscapes and wilderness areas. We must rethink and refeel our connections to and alter our relations with the land, basing them on an ethic of love and aesthetic beauty. Leopold argues that such a land-based ethics will alter human–land relations and the values that drive them. Environmental philosophy has largely been footnotes to Leopold's work, and justifiably so; his challenge to the utilitarian, anthropocentric ethics of his day remain forcefully pertinent for many environmental agrarians. Like Berry, Leopold would prefer culture begin by asking what is here in terms of a healthy landbase, followed by asking how we can fit our human culture and economy into this landscape, to become as he writes, "plain members and citizens" of the ecosystems within which we are embedded. The religious agrarians I study are indeed asking such questions.

Although many credit Leopold (along with Thoreau and John Muir as antecedents) for starting Western environmental ethics, there is another progenitor to environmental agrarian land ethics: Liberty Hyde Bailey. In 1915 this botanist and horticulturist wrote his treatise *The Holy Earth: Toward a New Environmental Ethic* that reflected his observations of increasing urbanization and pollution in an era that saw the first flourishing of industrial scientific approaches to farming. Within this context, he claims,

> Most of our difficulty with the earth lies in the effort to do what perhaps ought not to be done. . . . A good part of agriculture is to learn how to adapt one's work to nature, to fit the crop-scheme to the climate and to the soil and the facilities. To live in right relation with his natural conditions is one of the first lessons that a wise farmer or any other wise man learns. We are at pains to stress the importance of conduct; very well: conduct toward the earth is an essential part of it. (2009: 8–9)

This is one of the earliest environmental agrarian calls for a localized sustainable agriculture based on an ethic of restraint that equally calls for localized adaptation to and proper ethical relationship with the land. Bailey tellingly goes on to point out, based on his reading of the first lines of Genesis, that "The living creation is not exclusively man-centered: it is bio-centric" (ibid.: 23). Such an explicit call for a biocentric ethic predates Leopold's famous land ethic by more than two decades and laid the groundwork for today's environmental agrarian milieu. This includes Wendell Berry and Wes Jackson, who reference Bailey in their work.

Bailey prefigures other relevant and dominant environmental agrarian themes, writing that

> The farmer now raises a few prime products to sell, and then he buys his foods in the markets under label and tag; and he knows not who produced the materials, and he soon comes not to care. No thought of the seasons, and of the men and women who labored, of the place, of the kind of soil, of the special contribution of the native earth, come with the trademark of the brand. And so we all live mechanically, from shop to table, without contact, and irreverently. (2009: 65)

Although not as well-known as his fellow pre–Earth Day philosophers and scientists, Bailey nonetheless presages and echoes similar critiques of the emerging industrial agriculture paradigm as does Howard (1943 [1972]), King (2004), Balfour (2006), Faulkner (1944), and Steiner (2003); and of their descendants Berry (1977) and Fukuoka (2004); and today, Shiva (2000, 2016), Pollan (2006), Jackson (1985), Thompson (1995), and myriad others. When reflecting on the larger trajectory of the above strains of thought, it is evident that the environmental agrarian critique of the loss of a localized, land-based, sustainable farming culture has a century-old lineage. The reader should notice that this lineage includes an implicit call for spiritual and religious values to be placed on the land, as seen in the work of Bailey, Leopold, and Berry. Indeed, in the work of these three farmers, writers, and ethicists we see a move toward ethics and religious-like sensibilities (if not outright religious terminology and advocacy, where "nature" and land are central components of a new Earth-revering religious ontology) in their agrarian advocacy.

The sustainable farmlands of our country being worked by religious agrarians are becoming the loci of such a developing sacredness of the landscape. Many religious agrarians in the United States speak directly to this sentiment (see below), and share the insight of Native American scholar Vine Deloria Jr., who writes that "It remains for us to learn once again that we are a part of nature, not a transcendent species with no responsibilities to the natural world" (1994: 3). One goal of religious agrarianism is to act with ethical responsibility toward a divinely created world, where such ethics are part of a holistic worldview and give rise to holistic, sustainable farming practices (to be outlined later, and in consonance with what Beus and Dunlap [1990] observed in alternative agriculture circles where holism is a key value).

Other attempts to imbue landscapes with sacredness are well documented, and there exists a growing effort to use environmental ethics and holistic science as vehicles for forming human–land relations based on religious sensibilities (Callicott 1994; Rolston 1999; Harding 2006). Although I focus on examples of Jewish and Christian approaches in the United States to building on and re-forming sacred human–land relations in the context of sustainable farming regimes, it is important to note that these approaches are influenced by similar moves occurring within other strands of environmental and religious environmental ethics.[3] With the foregoing in mind, I briefly outline the centrality of the land in bioregional thought to show how local lifeways developed around local landscapes are central to building a sustainable culture, which is one of the key long-term goals of religious agrarians.

BIOREGIONAL THOUGHT ABOUT THE LAND

The "return to place" that is happening within the North American environmental milieu is cross-fertilizing with a nascent global, but especially US, ecological reformation. Such mingling of ethical values is contributing to changes in practices toward the land. What does it mean to say that the land is beginning to anchor values and practices within religious agrarian networks? How might religious individuals/groups respond to Berry's claim that "to farm is to be placed absolutely" (2005: 46)? What are long-term goals of land-based religious agrarian movements? Many of these goals are shared and articulated by permaculture, bioregional, and environmental agrarian movements. I have and continue to cover permaculture and environmental agrarian attitudes toward the land. Here I briefly explore the role land should play in perceived healthy human cultures as articulated from a bioregional perspective, then move on to investigating the views of land shared by the religious agrarians I interviewed, lived, and farmed with.

Bioregionalism can be summed up as the cultural, religious/spiritual, social, and economic reinhabitation of delineated ecological bioregions that occur on the Earth.[4] Part of this relocalization includes developing a lasting local food infrastructure. Aspects of bioregionalism can be found in the emerging sustainable food movement: for example, Slow Food, locavorism, and the hundred-mile diet, all of which support sourcing local foods. Proponents in these movements claim that they lead to the creation of a vibrant food culture dependent on regional land and farmscapes for sustainably sourced and crafted food. Such a "countercuisine *network*" (Trubek 2008: 115; emphasis added) is best defined by the concept of *terroir*, a French term that relates to food grown in a particular soil that then takes on attributes (especially flavors, tastes, and smells) of that soil. This concept has gained remarkable traction in United States food circles over the past fifteen years and contains within it a distinctive "set of values, practices, and aspirations" (ibid.: xv).[5]

In regard to her exploration of terroir and North American countercuisine food networks, author Amy Trubek writes, "According to this alternative model [n]etworks of people—farmers, chefs, and others—combine a quest for economic livelihood with the goal of a sustainable food system. But for these people the quest is also sensory: they want to create food of high quality. They are pursuing a business, a mission, *and* a craft" (Trubek 2008: 142; emphasis in original). Such a pursuit marries with stated ideals of environmental agrarianism and bioregionalism and is thus dependent on the terroir of place. This countercuisine network is not without its critics. Historically, most of those who are interested in such reinhabitation of foodscapes tend to be Caucasian, affluent, and well educated. This "baggage" of countercuisine networks became evident to Canadians Josee Johnston and Shyon Baumann, who found in their ethnography of the North American food counterculture that this foodscape movement contains "ideologies of distinction and democracy" (2010: xvi).[6]

Similar critiques extend to the bioregional movement in that most bioregionalists and bioregional philosophies have historically been generated from within educated Caucasian enclaves.[7] Many charge that bioregionalism is an ideal of luxury, in that if someone is poor, marginalized, and living in a toxic landscape, then staying put and building up a vibrant culture is not always a possibility or even a desire. In one respect, the call for New Monastics to move to the margins of empire can be seen as a Christian-based corrective to the perceived elitism of the foodie and bioregional movements. Koinonia's use of permaculture to become a demonstration plot for sustainable agriculture is part of this larger struggle to democratize and make practical living locally on the land (although, as noted, tensions exist within Koinonia about this part of their vision). Members of Hazon are also involved in democratizing alternative, localized food production as they contribute CSA shares to soup kitchens and help install sustainable backyard garden plots at centers for at-risk youth (see chapter 6).

Bioregionalism is also noteworthy in its approach to building a place-based culture because of its emphasis on creating and socially structuring place-based values. As Dan Flores writes, "Human cultures alter their places to shape them in accordance with their ideological visions, and in turn cultures are shaped by the power of their places" (1999: 52).[8] Sustainable farming practices, especially those based on environmental agrarian ethics and ideals, provide perfect opportunities for building and sustaining place-based local cultures. Sustainable farming practitioners tend to be conversant in local varieties of soils, flora, fauna, weather patterns, waterways, insects, and plant diseases and equally tend to build networks of information sharing with other farmers (Scherr and McNeely 2007). Such knowledge builds over time and is shared at farmers' markets, with customers who are CSA members, and through working relationships with retail stores that feature local artisan and sustainable food products. Those members of US religions who have religiously based and secular environmental concerns are actively seeking working relations with such farmers. Supporting sustainable agriculture practitioners becomes a way to put into practice religious and secular concerns for local landscapes and farmland.

I witnessed such concerns mobilize the religion-in-action of Jewish religious agrarians in the South with Atlanta-based Hazon members of Congregation Shearith Israel (CSI) and their desire to support Riverview Farm (which supplies CSI with their CSA produce). The bioregional ideal and concern has even been taken up by Nigel Savage, who stated in an interview, "I think the most complicated environmental issue for us has to do with place. I actually think it has to do with bio-regionalism" (Immergut 2008: 19). We see that concern for the local can be and is expressed in an amalgamation of environmental agrarian, permaculture, and bioregional ethics that mix with religious views of the land to create a lived practice of caring for local (farm) land.

TAKING A STAND IN, ON, AND OF THE LAND

One goal of this book is to explore how the development of agrarianism within the United States, especially environmental agrarianism, is influencing and shaping contemporary religious agrarianism. An articulate exploration of agrarianism and what it means for farming and politics is found in John Crowe Ransom's introduction to *I'll Take My Stand: The South and the Agrarian Tradition*, considered by some to "still [be] the best summary of agrarian principles versus the principles of industrialism" (Berry 2005: 43). This book, written collectively by the Twelve Southerners, has been hailed as a classic of literature and a bemoaned critique of the US proclivity to urbanize and march away from "traditional" agrarian values. It is also considered to be in "the family of works exploring what we might call *locational values*" (Heilman 1982: 102; emphasis added). However, the Twelve Southerners were rightly criticized for patriarchal tendencies, such that their appeal to Southern locational values applied to sections of the United States on a broad scale contained the seeds of its own cultural demise. Although an important and popular expression of US agrarianism, the patriarchal undertones of the Twelve Southerners made it hard for subsequent agrarians to take their vision wholeheartedly. This resulted in an important move by Berry: "[u]nlike the Agrarians, Berry did not look to history or ethnicity for the validation of values or for identity. Rather, *he found healing in attachment to place* . . . unlike the Agrarians, he was an ecologist" (Murphy 2001: 270–71; emphasis added).

What this means is that Berry's split from the agrarian past of the United States is made possible by the insights of evolutionary and ecological science. Therefore, the emergent US religious agrarianism outlined herein is a mix of agrarian history, ecologically informed agrarianism that is largely a criticism of fossil fuel–based Green Revolution industrial agriculture, and the ongoing ecological reformation. Together these attributes combine to form something that is unique in religion, agrarianism, and sustainable agriculture, respectively. Overall, religious agrarianism is based within a religious tradition (see Hazon within the larger Jewish tradition and Koinonia within the larger Christian and Protestant tradition), while also incorporating insights from ecology, soil science, and modern-day critiques of neoliberal economics and the industrialization of food systems. The latter critique is one of the original agrarian arguments of the Twelve Southerners and of other agrarians of the nineteenth and twentieth centuries. However, these historical agrarians could not foresee the environmental and social effects of the Green Revolution, nor of monoculture farming at scales undertaken today within a globalized and mechanized infrastructure of trading food items.

What is new and unique about religious agrarianism is precisely this mixing of religious beliefs, teachings, ethics, and practices in regard to the environment with the ecological agrarianism Berry laid out. For many religious agrarians I interacted with in the South and at food conferences, religion has something to say about the "local." The values and teachings religion gives about the local to the religious agrarians I met translates into practices geared toward embodying a deeper relationship with local farmlands, local farmers, and others (including animals, insects, soils, and soil microbes) within local sustainable food networks.

In his opening for the Twelve Southerners, Ransom opines that the forward-looking promises and slogans of industry and politics carries a subliminal message: "Do not allow yourself to feel homesick; form no such powerful attachments [to tradition and to place] that you will feel a pain in cutting them loose; prepare your spirit to be always on the move" (1962: 6). The spirit conveyed here is that Western life has not yet seen fit to ask Berry's questions about a local place. Instead, what I call the "tyranny of mobility" has allowed those in our country to pack up at the slightest whim so that a place can claim no fidelity to a person or vice versa. Agrarians, bioregionalists, and place-based cultures all find common fault with such a worldview. According to Ransom, the result of this deracinated tyranny is that "the latter-day societies have been seized—none quite so violently as our American one—with the strange idea that the human destiny is not to secure an honorable peace with nature, but to wage an unrelenting war on nature" (ibid.: 7). The antidote to this war is "to be moderate in our expectations of nature, and respectful; and out of so simple a thing as respect for the physical earth and its teeming life comes a primary joy, which is an inexhaustible source of arts and religions and philosophies" (ibid.: 9).

Here we see that most of the sustainable farming advocates, environmental agrarians, environmental ethicists, and religious environmentalists that I have been citing hold a view in common: cultivating a respect for the Earth and its ecosystems that includes the recognition that ecosystems have limits and that our survival is largely bound up in this recognition. This value of respect should then dictate practices—and not just any practice, but specific practices of sustainable farming—rooted in a local place, with implied respect based on insights from the environmental sciences. From such practice and respect for the teeming life of the Earth comes art, philosophy, and, most important for the purposes of my argument, religion. Also notice a change in values triggers a change in agricultural behaviors. I turn to the values and thus practices of US religious agrarians to see how they are putting their values about local place into practice and in the process creating lived networks in action founded on religious agrarian ideals and insights.

JEWISH VALUES AND PRACTICES ABOUT THE LAND

> Jews of America are urban, we're not rural. So there are a lot of Christian monastic communities who are farmers, but there is no Jewish analogue. We didn't start going back to the land until the '70s. Who has ever heard of a Jewish farmer? It's almost unheard of.
>
> —Beatrice Tischler

> *Hazan et ha-olain kulo b'tuvo b'chein b'chesed uv'rachamim* (who sustains the entire world with goodness, kindness, and compassion).
>
> —Birkat HaMazon, blessing after a meal
> (and painted on the wall at Isabella Freedman Center)[9]

My interviews and fieldwork with Jewish participants of Hazon evidence a strong ecological agrarian understanding of the importance of the local. This importance is central to creating an alternative to the globalized industrial agri-food system. Such understandings of the local are also grafted onto specific Jewish teachings and histories, leading to the creation of an emergent Jewish religious agrarianism that is taking root in various US (and Israeli) Jewish communities. As Beatrice Tischler notes, this emergent religious agrarianism dates back to the 1970s, making it a recent phenomenon in Judaism; it is becoming more pronounced and visible under the networking and advocacy of Hazon.

FRESH AND LOCAL

Many of the people I interviewed wanted to support local agriculture—sustainable or conventional, although across the board the preference was local and sustainable over local and conventional—because they believed that local agricultural products were superior in taste, flavor, and freshness. As Isaac Segal explained, he joined the Tuv Ha'Aretz CSA pick-up at CSI because of his desire for "better quality" produce that has more flavor. While being appreciative of Riverview Farms' organic farming practices, he is more interested in "just the taste. The taste of getting fresh stuff versus the stuff that's been sitting in a cold warehouse for a year before you buy it." Segal grows some of his own produce in his backyard and wants to provide his daughter with fresh food during the summer months.

Another Tuv Ha'Aretz participant at CSI is Beatrice Tischler, quoted in this section's epigraph. Like Segal, she prefers organic over conventional if given the choice but joined the CSA because her family "love[s] any kind of fresh fruits and vegetables." She also shares that she distinguishes a "taste difference" between organic and conventionally farmed foods and that she prefers the former; this is another reason she joined the CSA and is also why she is the "number one customer" at the Whole Foods across the street from her house.

A third Tuv Ha'Aretz participant is Naomi Eisenberg. She also happens to be on the board of CSI and was a participant in the job search that brought Rabbi Hillel to the synagogue. Eisenberg helped steer the creation of Tuv Ha'Aretz by supporting the initial work and vision of Dorothy Goldstein. She is also a committee member for the CSA partnership. Given this extensive involvement with the Tuv Ha'Aretz pickup, Eisenberg states, "as much as we were excited about having the fresh, organic food . . . it was really nice to say we're going to do our part to help the small farmer and support them." A driving concern for her is that people are losing their jobs, especially small family farmers. As such, she is pleased to be able to support Riverview Farm and enjoys visiting their farm and seeing the work they do. This support includes a strong interest in buying the grass-fed beef Riverview raises, but only if the cows could be made kosher by finding a local *shochet* (kosher ritual slaughterer) and processing plant (at the time of my work, such an arrangement had not yet been made or found).

These three reasons for joining Tuv Ha'Aretz reflect a desire to consume locally grown produce farmed in a sustainable way. All three CSA members feel that the produce tastes better, and all expressed the joy they receive when they get their weekly share and bring its bounty back to their kitchen. Such views reflect one of the guiding ideals of Hazon's Tuv Ha'Aretz program, in that "Tuv Ha'Aretz CSAs put the purchasing power of Jewish families behind local, sustainable farms. They deepen a connection to where our food comes from, and do so within a context of Jewish community and learning" (Savage and Stevenson 2009: 133).

Part of the context of Jewish community and learning is the weekly pickup. One of the pickups that I observed included the showing of the film *Fresh*, followed by an animated and articulate discussion.[10] Seven CSA members stayed for this discussion (six women and one man), and all seemed to know about Michael Pollan and Joel Salatin, two of the leading figures in US alternative agriculture circles. The movie spoke of many concerns associated with the local that were well received and echoed by the CSA members, but given the venue, it was only natural that audience members noticed that religious concerns were conspicuously not addressed in the movie. This is where the religious aspect of religious agrarianism comes into play—these Jews were watching this movie at a CSA pickup and agreed with and knew about many of the films messages. They enjoyed the film and agreed with its messages not only because of their own locavore, foodie, and sustainable agriculture concerns but expressly because of their Jewish identity and values.

This ability to have CSA pickups in a Jewish context and within a shared Jewish community makes the Tuv Ha'Aretz unique and compelling to many Jews and Hazon members. The pickup time also becomes a space for sharing recipes, memories of Jewish customs and meals, and discussions about food issues. It builds local community and networks Hazon members with local farms. During the film discussion, a few of the Tuv Ha'Aretz participants mentioned that the connection they feel when they pick up their produce is a spiritual connection to abundance and "something bigger." Whether this something bigger was God, the Jewish community, or the sustainable farming milieu was unclear, but some members articulated a sense of purpose that comes with joining the CSA and picking up their produce within a community setting.

This Jewish religious agrarian privileging and concern for the local was equally evident at the 2013 Food Conference. Daniel Infeld, senior program associate for Hazon, in welcoming all 216 participants from over twenty-two states, Canada, and Israel, pointed out that all of the food served in the kosher cafeteria came from local fields and farmers. According to Infeld, sourcing local foods helped the conference "embody our Jewish food values" of supporting the local. Furthermore, he pointed out that conference host Isabella Freedman has an educational organic garden, and all food scraps would be composted, allowing nutrients to stay local as well.

Privileging of the local was also shared by Janna Berger, the farm manager of Adamah farm, when she told participants in her workshop on food resiliency that "We are animals on Earth." For Berger, Jewish teachings and rituals speak to this emplaced evolutionary reality and give guidance on how to create local, sustainable farming in a specific place. Yigal Deutscher, Hazon's project manager for lobbying to reintroduce the *shmita* observance in contemporary Judaism, explained to a group of us on a permaculture walk[11] that when thinking about food and Judaism, we must "Always take the text of the Torah and sink into the land of its time—its food systems, climate systems, and ecosystems." For Deutscher, such sinking into the land can help answer the question of, in his words, "How do we develop societies and systems that are sustainable and permanent?" Deutscher is a leading voice in North American Jewish religious agrarianism, and a large part of this answer rests on fidelity to the local where such values informs him in thinking about Torah and putting such development into practice.

LOCAL HEALTH

Some of the Jewish agrarians I interacted with expanded on the "something bigger," citing a mixture of concern for local landbases and/or the health of local farming land, mixing this with a Jewish call to stewardship. Ruth Jacobs is executive director of B'nai Israel in Gainesville, Florida, and had just completed training with Hazon during their 2009 Food Conference on how to start up a CSA when I met with her. Passionate about sustainable food issues, Jacobs is a perfect exemplar of a religious agrarian operating within the Jewish tradition. This is seen in the following statement, which links her religious identity with agrarian concerns for local farmers:

> [What] is so incredible to me [is] that [for] every time that you eat, [we have] blessings . . . in Judaism. Whether we say "B'rai bre havetz," the fruit of the Earth, or the fruit of the tree, or "B'rai pre adamah," the fruit of the ground, there's this thread that draws whatever

> is on your plate back to the field and makes you think about it. And so while you might not be a farmer, Judaism dictates that you be aware of how that food was grown. And thus who was growing it, and where it was grown, and how it got from there to your plate. And that to me is really incredible, and I think it is something that people take for granted.

Jacobs echoes Segal, Eisenberg, and Tischler in that she enjoys fresher and tastier local food, and for her there's more: her tradition and how it is interpreted to speak about religious values that express concern for local farmers and the health of local farmlands and ecosystems. She felt driven to begin a Hazon-sponsored CSA partnership at her synagogue because "we know the environmental benefits of reducing our carbon footprint by eating local food; we know that it's fresher, we know that it's tastier. These are all things that we read in the Torah: protecting the Earth, caring for the land. So I think it embodies so many of the *mitzvoth* in that way. But I also think that it gives us a real opportunity. We live in a world where we are no longer farmers" and thus starting a CSA partnership is a way Jews can support local agriculture.

Moreover, as she states, her inspiration and vision is directly rooted in Torah. As she explained, "Just to have a connection to the land" via a CSA partnership helps bring Torah alive. For Jacobs the Torah is clear about duties of early Jewish farmers to leave paths in their fields, to give the land a Sabbath rest, and to not harvest corners of fields so the poor are able to have access to fresh food (these are all commandments being discussed via Hazon's Shmita project). Torah also speaks about crop rotation and how close and how far apart to plant different varieties of seed. Given this aspect of Jewish history, Jacobs explained to me, "I think it's an incredible teachable moment to have a farmer who we have a relationship with and to have a congregation-wide workday at the farm where we're out there in the soil, we're learning from them, we're seeing what they do. And then we can come back and we can study Torah. And we can say 'here's what we saw, here's what we didn't see,' and talk about it." It is in reinterpretations of tradition like this, where environmental agrarian concerns enter religious retrieval and reshape sustainable food-related rituals and activities, that Hazon and Jews in the United States are contributing to the development of contemporary religious agrarian practices and narratives.

The saliency of using lived religion and network tropes when analyzing religious agrarianism becomes evident when Jacobs states the following position: "with the force of what our congregation can do in terms of creating awareness and a support system to explore these issues and for becoming involved, the idealist in me says, 'Yes, absolutely. Religion is the vehicle for it.' I think that any organization, any community-based organization, is a vehicle for it. Judaism is a very natural pairing for this." Jacobs's active networks with Hazon and other progressive Jewish organizations supporting B'nai's efforts (she received a grant from a major Jewish grant-writing body, the Covenant Foundation, that allowed her to attend the food conference; she has also joined the Jewish Food Educators network to help bring Jewish food education to the synagogue, where the food director is going to start using Hazon's food curriculum with their Jewish Day School youth) are leading to religiously based values about sustainable farming being put into action. As she exclaims, religious partnerships and congregations are a "vehicle" and allow for values to be put into practice. To this end, B'nai has partnered with a local biodynamic farm, Sandhill Farm, to provide twenty members with a weekly pickup.

The most articulate response to queries about "the local" that I received came from the Jewish farmer and entrepreneur, Farmer D. This is a chosen name that appears on his business cards and that he uses when giving presentations on sustainable food issues (the "D" is short for his real name, Daron Jaffe).[12] Farmer D makes his living selling biodynamic composts, plants, and seedlings and by installing backyard gardens with these products at residences or various Jewish synagogues in Georgia.[13] I first heard about Farmer D in hushed, reverent tones during the 2009 Hazon Food Conference. He is a big name in the Jewish food movement, having begun the organic farm at Adamah, and he is a longtime participant in the Center on the Environment and Jewish Life.[14] This history helps Farmer D embody religious agrarianism from within the Jewish faith. He is also an exemplar of a lived networks approach to understanding social phenomena such as religious agrarianism, as he knows the owners of Riverview Farm. Farmer D and Riverview Farm are active in Georgia Organics (as is Katie Taylor of Koinonia, whom we meet later).

I was curious to learn more about Farmer D's journey and understanding of sustainable agriculture. *Sustainable agriculture* is a contested term, especially within the sustainable food movement, and given that Farmer D had once owned a 100+ acre organic farm that he started in college in Wisconsin, I began our interview by asking him to define *sustainable agriculture*. His response reflects his commitment to the biodynamic approach to farming begun in the early 1900s by Rudolph Steiner. For Farmer D, sustainable agriculture is one that embraces what he calls a biodynamics approach, which sees the farm as a holistic, living entity. Biodynamics is "a microcosmic approach to sustainability and it's a macrocosmic approach to sustainability. It's about how the farm relates to the outside world, and really how it relates to itself. How the farm feeds itself and provides for itself and the diversity and the balance that comes with biodynamic farms is about as sustainable as it gets." His experience as a Jewish religious agrarian led him to claim that "sustainable agriculture in a more generic term to me is more about farming, stewarding the land, using organic methods, regenerating the soil, not taking more than you can give back over the long term and not exploiting the land with chemicals."

This answer provides evidence of the crossover between religiously motivated sustainable farmers and ecological agrarian views about farming, where these merge to create a concern for the local (seeing the farm as a self-contained organism, influenced by Steiner's teachings) that is coupled with the basic tenets of organic farming. These tenets are traced back to Sir Albert Howard, who wrote in the very first paragraph of his classic treatise, "The maintenance of the fertility of the soil is the first condition of any permanent system of agriculture. In the ordinary processes of crop production fertility is steadily lost: its continuous restoration by means of manuring and soil management is therefore imperative" (1943: 1).

This response provides the nexus with religion, as Farmer D subsequently related, "I find my connection to the Earth and to agriculture, my connection to God through my relationship to the land." This sentiment is echoed in Midrash Ecclesiastes Rabbah 7:13, which states, "See my works, how fine and excellent they are! All that I created, I created for you. Reflect on this, and do not corrupt or desolate my world; for if you do, there will be no one to repair it after you."[15]

Farmer D shared his overall philosophy about sustainable farming and the importance of the local in a lengthy exchange that adroitly touches on many of the themes that can be found in US-based religious agrarianism. I began by inquiring about the statement regarding his relationship to the land and how this relationship provides a connection to God. I searched for more clarity on what he meant when he said "the land," asking him what this phrase meant and how important the land is to a healthy society, healthy politics, and healthy religion. Farmer D responded by saying that the land "is critical. If you look back, Gandhi speaks about how the way a society treats its animals is a reflection of the health of that society. I'm reading right now, Wendell Berry, Thoreau, Jane Austen. Modern-day agrarians that are trying to reconnect people back to what our true economy is around food." We see again Berry's importance in the development of religious agrarianism (it should be noted that members of Koinonia also actively read Berry's work), so in essence Berry has given contemporary agrarianism a lingua franca.

Farmer D continued by saying he is actively involved with Slow Food International. "When I first got into this, I was really curious about growing my own food. And then I started to realize, 'Wow, our Western civilization is founded on agriculture. And our agriculture is so industrial it's effecting the health and the way people live and the way they act and the impact on the environment and the houses we live in and the clothes we wear, and economies and labor. And family farms are going out of business left and right, like an endangered species.'" This epiphany is consistent with the critique of industrial farming put forth by sustainable agriculture advocates. Farmer D realized that the impacts of industrial farming influence "the social fabric, the whole social layer throughout agriculture, [and this] has a huge ripple effect."

Farmer D's motivation to enter into the world of sustainable agriculture was so he could participate in growing his own food. He realized that

> If people are disconnected from where their food comes from, they're cut off from the root of where they come from, which we're starting to get back to because of the economy. When everything is dependent upon fossil fuels and this cornucopia mentality and everything is an abundance and you can eat anything you want at any time of the year for cheap whatever, people don't have any kind of connection to the Earth and to [a] spirituality [of the Earth].

We see that besides familiarity with Berry's work, Farmer D also participates in lived networks with organizations like Slow Food International, maintains an articulate critique of industrial food systems consistent with critiques from leading sustainable agriculture advocates, evidences a concern for the social importance of local farming cultures, and most important practices and brings to his agrarianism a Jewish component of being "rooted" in the Earth via food. This rootedness is made possible by participating in local sustainable agriculture practices and signals a concern for local places, local health, and local economies that is rooted in religious agrarian values.

HAZON AND THE LOCAL

Religious agrarian viewpoints about the importance of the local—local farmland, soil health, community interaction and autonomy, and fresh food—were equally on evidence at Hazon's 2009 and 2013 Food Conferences. They are consistently present in a wide variety of monthly emails sent to members by Hazon staff. For example, on May 31, 2011, founder Nigel Savage included the following statement in Hazon's monthly email: "It's not a surprise that a new generation of Jewish people who care about health and sustainability have started to focus on these related cycles [of soil and society]. We learn from Wendell Berry that a society is as healthy as its soil, and a soil that isn't regenerated will decline and erode."[16] Notice how this is a new development in US Jewish practice, values, and identity (according to Savage) but also how Savage directly references Berry. Clearly Hazon is in agreement with Berry's agrarian views about the importance of local soil health for local social health; as such, this positions Hazon at the leading edge of Jewish religious agrarianism in the United States.

I witnessed similar sentiments at Hazon's 2009 Food Conference. At 5:30 pm on Thursday, December 24, 2009, the staff welcomed more than 500 participants to the conference. One of Hazon's executives, Mick Fine, welcomed those present by saying, "You are looking at the Jewish food movement. . . . We make up every place on the Jewish spectrum, and some of us aren't even on that spectrum . . . and we're all part of the Hazon family." Approximately 90 percent of the participants were at their first Hazon conference, and the entirety of the Hazon family were treated to a variety of presentations that directly related to the agrarian concern for the local. These included program tracks titled "Do-It-Yourself Food," "Food Systems and Sustainability," and "Jewish Tradition and Food: History and Culture." Specific talks ranged from "Food Entrepreneurs: How to Start and Run a Sustainable Business," "Decompose It Yourself: Compost for Commonfolk" (given by Farmer D), "Growing the Real Food Revolution on Campus," "Slow Money and Sustainable Food Systems," to "Community Involvement through Food."

Orthodox Rabbi Steve Greenberg, an executive director of Hazon, gave a talk titled "Can America Be the Promised Land?" Rabbi Greenberg pointed out that "Something [was] afoot with the way Jews are engaging with the land" by pointing out that most audience members did not live in the same community in which their grandparents did, such that "our attachment to a place on this planet is particularly fraught . . . today." For many Jews, Israel is the mythical, religious, and political land of attachment, leading him to ask if it is possible to be present to the land of North America while looking toward Zion? Greenberg argued that by looking toward Israel, Jews do not pay attention to the Earth where they live, and Hazon is helping remedy this—thus, something is "afoot" with Jews today. Yet he also pointed out that the original exile and homelessness of Jews makes them the original "canaries in the coalmine" for all transient people of the postmodern world of metropoles and peripheries. The call for Jewish renewal by returning to the Promised Land is a template shared by the US agrarian tradition, which according to Greenberg uses the same language: people need soil and land where they can build their own cultural identity such that the land nourishes the people and vice versa.

For Rabbi Greenberg, Jews have needed this for 2,000 years, and now most US citizens need it, too.

With this presentation Rabbi Greenberg made the most explicit link between the bourgeoning Jewish food movement, the Jewish history of diaspora and exile, and US agrarianism that I encountered at the 2009 conference. This link was made clear when he handed out a flyer that had various quotes from Berry on it plus quotes from Ellen Davis's agrarian reading of the Hebrew Bible. Clearly, many within Hazon's leadership of are making explicit links with ecological agrarian philosophy and they are comfortable with sharing the strengths of the agrarian vision with Judaism's storied history about human–land relations.

Another well-attended talk at the 2009 conference was titled "The Vegetable Monologues: Jewish Female Farmers." Four female Jewish farmers spoke about their experience working on sustainable farms and/or starting businesses that sold organic produce.[17] The talk had an overflow of people sitting in the aisles, and it generated quite a bit of dialogue and heartfelt questions between the presenters and audience. The four panelists expressed the desire to create healthy, local food systems and shared that this desire is what drove their farm practices. One farmer makes goat cheese with her husband and pointed out that "the methods of production" matter because they have long-term impacts on food and water systems. Another presenter, Anna Hanau, was then the farm manager at Adamah, and she and her husband were in the process of starting a kosher grass-fed beef slaughterhouse.[18] She went to Jewish Theological Seminary as an undergraduate and double-majored in biblical and environmental studies. After graduating, she met Nigel Savage and declared, "It had never occurred to me before that there are Jews that find God in the woods and sunbeams," although she came to realize that "our tradition speaks to this." At the time of the conference Hanau had worked for Hazon for three years and was instrumental in beginning the CSA program. Hanau spoke of how she feels connected to her Jewish roots, stating, "Our ancestors started [out] in awe in the natural world." All the presenters shared how their experience as farmers has made them appreciate the beauty of their local ecosystems and farmlands, discussed the very real responsibility to these landscapes that comes with being sustainable farmers working within natural systems, mentioned how important it is to support local farmers so they can continue land stewardship practices while making a living wage, and agreed that the Jewish tradition speaks to these concerns. Indeed, these various themes were central to the conference as a whole.

Hazon's *Food for Thought* touches on these themes, as well. The authors write that "Jews have two different stories about place: we have lived in nearly every country in the world, but our tradition is focused on the Land of Israel"

(Savage and Stevenson 2009: 81). This tension was one of the key themes that guided Rabbi Greenberg's talk. The authors continue, stating that most Jews operate under three sometimes competing images: the wandering Jew, the pull of Eretz Yisrael, and being part of local and regional food traditions (ibid.: 83). Although "the Jewish people entered history as an indigenous people—with a direct relationship to the food native to their land" (ibid.: 84), the authors feel that "Given that much of our food sometimes seems placeless, we think that it's a positive value to consciously eat locally" (ibid.: 86).

Last, there is one more level of the local I encountered with Jewish religious agrarians: a local interpretation of Torah, or God's instructions for living, that speaks to local soil, health, and land needs of Jews living in a local place. This sentiment was most eloquently expressed to me by Gavin Sanders, an employee of Tekumah, "the first and only Jewish organization dedicated to teaching the holy trades to any Jew interested in learning—regardless of gender, sexual orientation, or religious observance."[19] In 2013 I arrived to Isabella Freedman early and helped Hazon staff set up chairs and organize the retreat center for the conference. During this time I had the chance to work with Sanders. I asked what had brought him from a Conservative rabbinical program in Los Angeles to the conference, and he shared that his wife had been to the Hazon Intentional Community Conference held in Baltimore in 2013 and they had been to Israel to visit ecovillages and sustainable kibbutzim. These experiences influenced his path as a Jew, and he was very interested in learning more. When queried about what and who is driving Hazon, younger Jews (he and I and others setting up for the conference all were in our twenties and thirties) or older Jews, he quipped, as we carried chairs through the snow to another room, "the older ones are old hippies with more money, and the younger ones do all the work." When I asked what he was studying, he said his goal is to become a *shochet* (ritual slaughterer), so that he could teach nonrabbis how to do ritual slaughter. In his words, his goal is local Jewish revitalization, where he can help "democratize holy trades" that historically have been controlled by the ultra-Orthodox, especially the trades of scribe and kosher slaughterer. Sanders thus speaks to another emerging element in Jewish religious agrarianism in the United States regarding the local: localizing and democratizing Jewish ritual practice, especially as this relates to food issues.

CHRISTIAN VALUES AND PRACTICES ABOUT THE LAND

Koinonia members come from a different religious background and live in a different context than the Hazon members in Atlanta and Gainesville and those at the Food Conferences. Their reasons for being exemplars of religious agrarian concerns for the local are theologically and culturally different in a religious sense. From an environmental standpoint, they are similar and have significant overlap. Furthermore, those working the land at Koinonia are by default intimately engaged with the local, since they are managing and interacting with land that is literally their home and backyard: the buildings that make up the campus are surrounded by the community's pecan orchards, blueberries, muscadine grapes, pasture fields, organic vegetable garden, and permaculture fields. Simply by walking from a residential building to the communal kitchen or across the street to the chapel emplaces a person into the middle of Koinonia's nearly 500-acre ecosystem. For those who manage, care for, and run this landscape, concern for the local is always front and center.

THE SACREDNESS OF LOCAL CREATION

One of the matriarchs of Koinonia (who has since moved to a retirement home to be closer to her children) was Sylvia Castle, wife of Dave Castle.[20] She always greeted visitors with a smile, and her exuberance, humbleness, and commitment to Quaker pacifism belied her age and slight build. Castle strongly supported the community's vision to embrace permaculture, a turn she saw as part of God's vision for Koinonia. As she stated, "Well, I mean we're all one, with all of creation. So the Kingdom ought to look like 'Yes, we're all connected and caring about-for one another and with one another and that the earth would be bountiful in a way that God intended.'" With this view of creation, Castle believes the human role is to be stewards, as she "think[s] it's all God's creation and I see it as honoring God, not replacing God [to care for God's creation]." Such a role and theological view motivates Castle to support the community's decision to care for its land in a holistic way. For her and others, Koinonia's land is a place to demonstrate the kingdom of God to visitors so they can see other ways of interacting with the local and of growing food in a sustainable way.

Another community member, Paul Robinson, heard about Koinonia because of his Christian-based interest in permaculture and desire to lessen his impact on God's creation.[21] As a self-described "child of God," Robinson sees Koinonia as a perfect place to birth a new vision of local human–nature relations. As he described, he is motivated to care about the local because he believes the Earth is God's creation, and species extinction, human suffering, and overconsumption are all topics about which he is concerned. Robinson moved to Koinonia because at Koinonia he is able to act on these religiously motivated concerns "locally and in the context of where I am right now. At least in the community I can act on it right here and be changing my lifestyle and helping to change the lifestyle of *this community* and then I believe that other people will be affected by that lifestyle change." Notice that Robinson feels he can live out his values at Koinonia, and that this latter belief reflects the community's standing as a "demonstration plot" visited by thousands of people each year. Robinson further explained that Koinonia is part of various lived networks so that the community "has a lot of connection across the country and across the world . . . Koinonia has a large influence across the world. And [if] many of those people get word that we're starting to do things differently it might trigger something in their minds."

This part of our interview captures a Protestant religious agrarian concern for the local. We see that Robinson's values (his concern) are based on a theological concept of a holy creation. Becoming educated about how this creation is being treated motivated him to live at Koinonia where he can put into practice permaculture principles on a local scale in a living community, with the hope that Koinonia's fields will motivate others involved in lived networks with Koinonia.

Robinson's roommate at Koinonia is Ben Peters, and together they are two of the most committed members in terms of actively pushing other community members to embrace the community's sustainability vision. They work on the garden crew, have a compost toilet in their house, and do not run the air conditioner despite 100-degree summer temperatures because it uses electricity from a power grid. Peters came to Koinonia in his early twenties after working on faith-based justice issues in Philadelphia. His inspirations include Leo Tolstoy, Daniel Quinn, Derrick Jensen, and Jesus's ethical teachings in the Sermon on the Mount.

When asked to define organic farming, Peters responded, "I guess in the best sense of the word it's growing food in a way that regenerates the landbase on which you're growing it. And the worst way it's a tool used by corporate farmers to make more money for themselves" (notice the similarity with Farmer D's biodynamic approach to sustainable farming). This shows Peters's knowledge of the basic claim of organic farming (and ecological agrarianism), as well as insights into the various debates over the contested meaning of organics and the perceived takeover of organics by larger corporate interests. When asked about Koinonia's commitment to sustainability and the reason for this commitment, Peters answered, "part of the Kingdom of God is healthy relationships between us and the Earth, which is how I view it." This implies that his support of sustainability is based in a theological understanding of human–Earth–divine relations. This understanding led Peters to move to Koinonia so he could practice his religious values and convictions and act on his desire to generate healthy and local sustainable food. As he puts it, "It'd be great if like every single thing we eat here would be from the garden [and] maybe even trading with some people" who are growing beans and grains (as protein for vegetarians like Peters). Such a sentiment expresses the agrarian concern for local foods, local community, and local responsibility. For Peters, these concerns are motivated by and couched within his Protestant conceptions of the divine and by membership in Koinonia.

A similar perspective on the Christian religious agrarian belief about the sacredness of the local is found in Stephanie Paterson. Paterson grew up home-schooled in an urban punk household and found Jesus in her late teens at a punk show for Christians. Since this experience, she has traveled the country, lived at a Catholic Worker House in Los Angeles, and worked in the organic gardens of an all-girls school in the mountains of California. I met Paterson as she was visiting Koinonia for three weeks to see if she wanted to become a community member.[22] I found her to be creative, vibrant, and full of passionate ideals motivated by her love for God and the Earth. She feels people in the United States "are becoming less and less intimate with the Earth. Because of technology and just the way our culture is progressing and that's really dangerous to me because it seems like when people lived more intimately with the Earth they weren't abusing it." In this respect, Paterson mirrors Robinson's and Peters's sentiments that smaller scale tribal and community settings centered on sustainable relations with the Earth are the ideal form of human community. For them, Koinonia precisely fills that role and does so in a way that allows them to honor God, God's creation, and fellowship with other Christians.

This sentiment was captured in a conversation with Paterson after we spent a day picking blueberries. During our conversation, she revealed that "working in the garden here I have realized that like that's a bigger part of who I am than I ever realized." When asked for clarity, she continued, "working with permaculture . . . just working in the garden and seeing the way they're handling the plants . . . it's just everything is so sacred here. They treat the Earth like such a sacred thing and I've just never really experienced people loving the Earth that . . . that's a big reason why I've decided to stay here. Not just because of the community but because of the connection with the land that I've found here." She shared that this connection helped bring Christianity alive. She explained this feeling by relating, "when I read [in the Gospels] love your neighbor as yourself, I think we had better take care of each other, we can't sit around while other people are poor. And love God, and to me God is the Earth, so I mean, it's my own personal interpretation, but I think it should definitely be a fundamental value of all Christians and all spiritual people . . . the way I feel out in the garden, I can't imagine how not everybody doesn't feel that way." I inquired if this feeling was always positive and beneficial, given that at the moment it was 100 degrees and humid and we were working near red ants. Undaunted, Paterson replied, "I'm covered in ant bites . . . you know, it drives me crazy when people say 'I'm an environmentalist' or 'I care about the environment.' You *are* the environment, come on! This is us, this is who we are, this is what we're meant to be, we're a part of this; we are part of the ecosystem."

Our conversation captured an almost pantheist view of human–Earth relations, one that sees the Earth as being divine and humans as being part of this sacred Earth. The environmental values Paterson developed based on her religious beliefs led her to want to use permaculture to build a local culture of honoring the Earth through combined prayer and action. Her religious agrarianism even accepts the ant and mosquito bites that entail to being part of an emplaced, local ecosystem. If operating under an industrial agriculture model, the same ants and mosquitoes would be eradicated by using various toxic chemicals that would then leach into local soils and waterways. The religious agrarians of Koinonia have a different view of these same insects and of the soils and waters found on their campus—they are all sacred because they are God's creation. It becomes their task to care for them by generating healthy, sustainable, local farming practices as outlined in their vision statements.

DESIGNING LOCALITY

The generation of religious agrarian practices at Koinonia is based on the permaculture ideal as originally developed by Mollison and Holmgren, which was originally introduced to the community by Jim Everett. Katie Taylor is a steward of the community who is married to the lead farmer/gardener, Joel Taylor. Together they have two daughters and live on campus near the cafeteria. As stewards, they have pledged to live in the community for the rest of their lives. In their early thirties, they are active in worship and helping Koinonia embody its mission statement. Taylor is in charge of hospitality and gives tours to the various groups who visit through the year. She has also received training in permaculture design and helps run the permaculture design courses that Koinonia holds one or two times a year, including the training I attended in February 2010.

When giving tours, Taylor tells visitors that

> Permaculture, the most succinct definition I have for it is, it's a design system for creating sustainable human settlements that work

> in harmony with nature and work to produce an abundance of everything you need. They talk a lot about inputs and outputs and so the idea is that you eliminate or minimize the inputs, which is something you have to bring in from outside, like fertilizer, seed, [and] fuel, [so you] try to reduce your reliance on those things through seed saving, and soil building with things you have on site rather than going and buying mulch. . . . And then the outputs, what most people would call waste, we try to reconsume as many outputs as possible.

In this way, Koinonia is committed to reducing their impact on local and international levels of trade and commerce (including especially in food commodities) by creating as sustainable a system as possible.

In fact, the intentional design aspect of permaculture is what piqued Paul Robinson's interest and is largely what brought him to Koinonia. For Robinson permaculture "is about design and being able to realize that we can design our lives and our surroundings so that they're all in harmony and providing for our own needs and being mutually beneficial for each part of the system that we can design. So that methodology that was kind of laid out in permaculture kind of appealed to my engineering background and desires because engineering is design, so that attracted me a lot." Here he references his undergraduate studies in engineering and how he can apply those skills to permaculture design to make harmonious systems where humans are an integrated part of a local landscape rather than a species that colonizes most of the available biomass. Robinson found Koinonia online as he was searching for permaculture communities, and the Christian identity convinced him to join.

DEMONSTRATING PERMACULTURE ON A LOCAL SCALE

This religious agrarian mix of permaculture and Christianity is quickly becoming one of Koinonia's main ministries. As Taylor explains, "I think that permaculture is one of the next big things that we're going to . . . gift to the world." This is in reference to the other gifts Koinonia has demonstrated: racial fellowship, Habitat for Humanity, civil rights campaigning, and peace activism. For Taylor this gift is important because it is moving sustainable food concerns out of "hippie pagan Earth-loving, non-Christian circles" and bringing them into modern Christian consciousness. She argues that such work is important for Koinonia to carry out because people within "the Christian faith are starting to look for that. People in the Church are getting disillusioned and they're looking for things and connecting them with permaculture, which is . . . just a way of looking at the world, and it can be applied to anything." Koinonia is applying permaculture to their local land and developing Christian religious agrarian values and practices. This work is important for Taylor and others because "I guess I see it that's what we are, just drawing people together" from various Christian subtraditions and exposing them to permaculture by providing trainings. Such work led her to claim that "For me, the reason I think I was able to grow a relationship with God, here, versus all the other places in my life [is] because I'm at peace now [and I am part of this larger role we're playing for God]." Taylor's religious values and beliefs marry with the vision of Koinonia so that she feels that the community has a role to play in shaping the future of Christian–Earth relations so these become more sustainable and in harmony with God.

My interview with Taylor uncovered her views of nature and how her understanding of sustainable farming practices influences her lifestyle choices and thus concepts of proper human–nature relationships in a local setting via food choice. When asked about the role nature plays in her religious beliefs, she asserted that nature "just inspires awe in me. The way things work in harmony with each other. . . . It already *is* the right system. And the way that like life and death cycles through it is really big for me." When asked if she sees this system as being God's creation, Taylor emphatically responded by saying, "Oh yes . . . and so I guess for me, I do feel very called to just be in relationship with nature as well. That, the Earth is not something separate from me that I am allowed to use, it's something that I need to work to be in harmony with." Taylor admits that living in "harmony" with nature is a challenge, and it might not be possible. However, by living at Koinonia, she is able to grapple with putting into practice her desire to live in harmony with God's creation. As she explains, "For me there's still some unanswered questions and some things I don't quite understand . . . I was vegetarian for ten years and I've started eating meat again. And we raise animals on the farm for meat. And I still sometimes am like, 'Is that right or wrong?' I'm not really sure, but this is what I feel like I'm supposed to be doing now, so I'm going to go with it." One reason she is fine with eating the meat of animals raised by Koinonia is because "I know it's better for the Earth and better for those animals in the long run for them to be raised this way," a view espoused by Naf and Anna Hanau, who began Grow and Behold. Taylor eats Koinonia's locally raised meat "because it's better than growing a whole monocrop field of soybeans and spraying chemicals on it or even organic soybeans, like there's a lot of input and a lot of fuel and a lot of stuff that goes into that. And it's not the best thing for the soil and it's not the best thing for people and you know importing produce from Brazil is not a good practice."

“CHICKEN MOBILES” AT KOINONIA WITH COVER CROP FOR BUILDING SOIL HEALTH. Photo by the author.

PERMACULTURE-DESIGNED GARDEN PLOT AT KOINONIA.
Photo by the author.

PERMACULTURE-DESIGNED FIELD AT KOINONIA. Photo by the author.

SIGN FOR CSA MEMBERS AT CONGREGATION SHEARITH ISRAEL.
Photo by the author.

CSA BOXES AT CONGREGATION SHEARITH ISRAEL WAITING TO BE PICKED UP BY TUV HA'ARETZ MEMBERS. Photo by the author.

CSA MEMBERS TOUR RIVERVIEW FARMS. Photo by the author.

We see in these reflections that Taylor spends a lot of time thinking about her food choices and how these choices influence various ecosystems. By practicing permaculture and consuming the products grown locally, Taylor and other members are able to grapple with, develop, and practice religious agrarian ideals that challenge the industrial farming model. However, these ideals are part of a larger Christian worldview that sees the land as a sacred creation and sees Koinonia as having a role to play in spreading the religious agrarian "gospel" to other Christians. In this respect, the religious agrarian practices of Koinonia are shaped, informed, and inspired by religious values. Last, Koinonia provides the seedbed where these religious agrarians can cultivate their sustainable food values and, even more, put them in practice.

One such person who was exposed to Koinonia's Christian-influenced religious agrarian practice of permaculture is Emily Hoffman, an attendee at the February 2010 permaculture training. Hoffman is a member of one of Koinonia's sister communities, Jubilee Partners, and her participation in the training provides another example of a lived network.[23] This is because Jubilee and Koinonia are both New Monastic communities and experience fellowship together throughout the year. Jubilee was started by a former Koinonia member. Both communities are committed to peace and immigrant rights, and the latter is the focal point of Jubilee's ministry. Immigrants from war-ravaged countries are given homes at Jubilee, where they are taught English and are given land so they may start their own gardens. Jubilee has a noncertified organic garden on site, where Hoffman works. This background led her to attend the permaculture training so that she could help Jubilee transition to a permaculture-designed campus.

When asked about what got her interested in supporting local, sustainable agriculture, she responded, "I guess for me at some point ten or twelve years ago a light bulb [went off as] I was tending the first garden that I had and was watching the food grow and eating it and just realizing this is good work to do. To grow food to put in one's body, it makes a lot of sense." This is a similar epiphany experienced by Farmer D, showing that the embodied, religion-from-below approach of lived religion is able to capture the emergence of religious agrarian sensibilities.

When asked if "good work" was to be interpreted biblically, Hoffman stated, "it felt like, obviously since the beginning, Adam and Eve tended to their food. It made sense because if we're supposed to nourish ourselves with some sort of food we may as well make sure it comes from a good source and tend to it and not be disconnected from it in anyway, so the connection I guess felt like something that God sees as good for us." Similar sentiments are expressed by Paterson and Taylor, so for them, fresh "fruits" of agriculture are activities that are pleasing to God. For Hoffman this divine connection with food provides a connection with the land and the seasons. As she explained, "I guess if you need to put food in your body, why would you not be connected to it?" For Hoffman and religious agrarians broadly, this connection not only is with a theocentric creation but is made readily available by the participation in localized, sustainable agriculture practices.

FARMING PRACTICES THAT BENEFIT THE LOCAL

Members of the Hazon and Koinonia communities value local landscapes, local farmland, local farmers, local economies, and local produce, leading to practices of localized sustainable agriculture. At Hazon the concern for the local incorporates themes such as taste and freshness, creating an agrarian-infused kosher that reflects Jewish diets and ethics in the modern world. There is an aversion to using what are perceived as dangerous chemicals and poisons on local farmlands and a desire to support local farmers. At Koinonia and in the lived networks of its members, the concern for the local incorporates themes of self-sufficiency, not using what they perceive to be toxic chemicals on their own farmlands (see next chapter), a desire to live in harmony with God's creation, and their view that they have a calling to "demonstrate" the benefits of permaculture to visiting Christians and non-Christians alike. Given their shared religious agrarian concerns for the local, I briefly summarize local practices of agrarian farming that both groups embody (or attempt to embody) before moving my exploration in the next chapter to the religious agrarian theme of "health."

HAZON

Some religious agrarians involved with Hazon in Atlanta grow some fruit and vegetables and herbs in their backyard, but most fit the description of being "urban agrarians" (Donahue 2001; Orr 2001). As such, they support local farmers and businesses at farmers' markets and by being members of CSAs. Farmer D has his own local business, and B'nai Israel in Gainesville partners with a local biodynamic farm while entertaining ideas of growing food on their campus. However, most of the people I encountered in Atlanta support sustainable farming practices by entering a CSA partnership with Riverview Farms, a family-owned farm in the northwest foothills of Georgia.[24] I visited

Riverview Farms on two occasions—once to get a tour from the owners, and the other time on a field trip to the farm with CSI members. The latter visit included a picnic followed by a farm tour so that the members could learn more about from where their food came. Quite a few families attended this and some brought children—not a small feat as Riverview is one and a half hours away from the synagogue.

Riverview is bounded by the Coosawattee River, and nearby is Carters Lake, an old Cherokee center. The farm is surrounded by pine plantations and meandering creeks and contains fertile valley soil of the Appalachian foothills (visible in the distance). There are hay and alfalfa farms, corn monocrops, cattle farms, small rural towns, trucks with hunting dogs in back, and a plethora of Baptist churches in the vicinity. This immediate context means that Riverview is an oasis of progressive organic farming practices in a sea of rural northwest Georgia politics, farming, and a cultural mix of the two. This part of the state is similar to the rural poverty of Americus—rusty trailers, churches, and abused land seem to go hand in hand with a slower pace of life compared to that of Augusta, Atlanta, and Athens, although the demographics near Riverview tend to be predominantly Caucasian.

Riverview is farmed by Tricia and Ron Daniels, a husband-and-wife team who live on the property with their young child. Ron Daniels's parents are the owners, who used to farm the land by growing monoculture row crops of soybean and corn and raising hogs. Unfortunately, these practices led to a significant amount of debt; by the late 1990s Mr. Daniels was in a cycle of increasing debt coupled with increasing use of chemicals. In 2000, the younger Daniels, along with his twin brother, Shannon, moved back to the farm (Tricia and Ron met at the University of Georgia, where they were chemists, with Tricia hailing from Americus and Ron saying that the Agriculture School of UGA was simply a funneling device for agribusiness interests) and began to transition the farm to organic. They are now certified organic, with Tricia serving on the board of Georgia Organics. They also have resident interns because part of their goal is to teach a new generation of farmers about how to run a successful small organic farm operation. Ron's father is on board because for the first time in years, the farm is financially solvent, and their customer base expands each year.

According to their website, they are "One of the oldest, and largest, certified organic farms in the state, [where] our goal is to work with the environment to enliven the farm through the mutually supported functions of soil, grasses, and animals as we produce organic vegetables and meats. With detailed attention to soil quality, animal husbandry, grass production, and crop diversity, our methods surpass organic certification requirements. Good for the land, good for the animals, and good for people too."[25] On their 200 acres, the Daniels raise 150 cows and butcher 8 to 10 hogs a week for their meat CSA. They use a clover-rye-vetch cover crop to build nitrogen and soil humus, grow corn and soybean to feed the hogs, let the cattle range free and feed on grass, store the waste run-off from the hogs in a pond and use it to water the fields twice a year, and grow collards, radishes, turnips, garlic, cilantro, black beans, tomatoes, squash, carrots, strawberries, and a variety of other vegetables. In all, Riverview has fifteen acres in rotation with various vegetables and the rest of the acreage is used growing grains and letting animals pasture, with Shannon Daniels using some of the corn to grind on site and sell around the region as organic grits and corn meal.

The rabbi was one of their first CSA members, and this early relationship led to Dorothy Goldstein tracking Tricia Daniels down when the original partnership CSI had made with Tayl Organics ended. "If a farm can't make money, you can't farm," as Daniels points out, so it is vital to their success to have the partnership with CSI. When asked if there were tensions with anyone because religion was part of the partnership, Daniels responded "not at all," and added that "We don't go to church. We're here on our knees every day. This is church." Overall, Riverview has 350 CSA accounts, just developed a "veggie mobile" that brings their produce into Atlanta where it is sold out of the truck, and serves two farmers' markets in Atlanta.

By participating in a CSA partnership with Riverview, Hazon members create lived networks with local farmers (organic apples and eggs come from other farmers and Riverview adds these into box shares), support the financial stability of a family farm, and indirectly support the organic farming practices and preservation of 200 acres. The latter link was one reason a CSA member brought her young child to Riverview on the Hazon picnic so that, in her words, her child will not be "an ignorant city kid." The ability to visit a working, sustainable family farm and get fresh produce from that farm was an important reason this member joined the CSA. In the process she felt she contributed to creating a vibrant, strong, nourishing, and sustainable local food culture.

KOINONIA

Koinonia has a long history of farming, the results of which can be visibly read on the landscape. The most prevalent visual aspect of this history are the large pecan groves that dominate parts of the campus. These provide the main revenue for the community, and because they are farmed conventionally, these groves are now an anomaly compared with the rest of the sustainable farming practices instituted by Joel Taylor under the community's sustainability vision.

It must be noted that Koinonia members, working with researchers from the University of Georgia and Georgia Organics, are attempting to transition some of their pecan trees to organic production. The biggest problem is that mites enter into the fruit right at formation so that the nuts never fully develop. Another fungus attacks the leaves, killing them and thus affecting pecan production. Together, both are major threats to pecan growers everywhere and the only successful defense has been to use industrial fungicides and pesticides to get nuts to reach maturity. Taylor shared with me that as of the 2014 growing season, "We are still using the biological management practices for the pecans. It's going all right. Our twenty-five-acre Elliot orchard is coming along quite nicely, but the Desirables (which are most of the rest of the acreage) are struggling with a fungus called scab. But we had a good crop last year, and we are committed to using organic/ chemical-free pecans for the future, as far as I can tell."[26]

The rest of Koinonia's farming practices are a paean to permaculture and sustainable farming practices. They have you-pick organic blueberries and muscadine grapes; they use pecan shells, muck out the chicken coops, and use cardboard to build soil; use cover crops to build nitrogen and soil; use rotational grazing with their small but growing cow herd (along with sheep, goats, and hogs) which they are building up to eventually make a meat CSA; provide milk and meat for those in the community from their goats (and meat from pigs, turkeys, geese, and rabbits, who also supply compost); have bee hives for pollination and honey products; have pear, apple, plum, and other fruit trees; and grow vegetables and fruits in the garden, including strawberries, greens, squash, corn, beans, rhubarb, tomatoes, and a variety of other crops, all of which end up in the community kitchen. Last, a few community members are committed dumpster divers, who go to Americus a few days a week to look for discarded food which they bring back to feed the animals and compost. Community members see this practice as a way to take waste out of landfills and to use local resources to build the soil and animal health of the farm.

The fields adjacent to the buildings on campus are designed along principles of permaculture design, with the goal being a mixed edible forest that has annual food crops while also providing habitat to chicken, geese, ducks, and pigs, all of which will compost and turn up the soil. Swales have been built in some of these fields to catch and retain groundwater, and a chicken mobile moves throughout the fields so that the chickens fertilize targeted areas. Across from the buildings and adjacent to the main pecan groves is a fenced-in eighty-acre pasture that holds sheep, goats, and cows. Koinonia members engage in "efforts to build a herd of the endangered Pineywood cattle, a breed that is naturalized to our bioregion but are nearly extinct since the introduction of European breeds and excessive crossbreeding. Our hope is that the re-introduction of Pineywoods will nourish our land and help to create a stable ecosystem for other endangered species that were once native to this area."[27]

In efforts to further their move toward local sustainability while being a demonstration plot for God, Joel Taylor and Lisa Jones attended a three-day workshop at a farm in Tennessee devoted to holistic resource management. This workshop was described in one of Koinonia's monthly email newsletters as, "a systems thinking approach to managing land resources and increasing production, while at the same time building biodiversity and improving the quality of life for those who use it. Holistic Management . . . provide[s] a wonderful boost of encouragement to our efforts as we continue learning how to heal our land and to grow an abundance of healthy food."[28] Koinonia added their own emphasis to the final part of this email, signaling their continued commitment to stewardship and caring for their local campus. This commitment is thoroughly shaped by the institutional history of the community, including the values put into practice by various members, past and present. However, this commitment also takes seriously the goals and standards of sustainable agriculture and environmental agrarianism, including the healing of human–nature interactions. Yet these goals and embodied commitments occur in a clearly articulated Christian context where Christian values motivate sustainable farming practices. Because of this, Koinonia is an exemplar of religious agrarian concerns about locality.

THE LOCAL: CODA

In Paul Thompson's most recent book, part of the series on New Agrarianism edited by Norman Wirzba, he asserts that "A philosophy of agriculture—and, by extension, sustainability—is thus appropriately attentive to dimensions of place or locale, on the one hand, and local or regional history, on the other . . . belief in and attentiveness to characteristics that make a given place 'special' are hallmarks of agrarian philosophy" (2010: 9). The Jews of Hazon and the Christians of Koinonia are working, in unique and similar ways, to build agriculture and sustainability at local levels that are bounded within religious and regional histories and landscapes. Their work situates them respectively at the vanguard of developing a domestic US religious agrarianism.

It is important that this journey does not degenerate into either land fascism or political or religious posturing. Rather, the development of a religious agrarianism is an iterative process that requires community, the willingness to learn, and an engagement with politics (see chapter 6). However, this process is clouded by outmigration (from Koinonia, synagogues, CSA membership, or religious identity in general), by absentee ownership of lands or the practice of farming methods that are perceived to be inimical to sustainable farming practices (for example, the industrial methods utilized on lands that bound Riverview and Koinonia, respectively), and by the lack of transparency in laws (especially the US Farm Bill) and/or the resistance regional laws put in place that limit governmental (and economic) support of practices that support the health of the local.

Most organic farmers I've met, interviewed, or worked with are progressive in terms of their critique of industrial agriculture and what they perceive to be governmental complicity in subsidizing and encouraging deleterious industrial farming practices. Yet they are conservative in that they want to farm the right way and want to conserve local lands and the best of local cultures. Many sustainable agriculture proponents I met are critical of industrial progress (but use it as needed—for example, they may buy new tractors and organic-certified chemicals and use new labor-saving technologies) and therefore would join with the New Monastics in criticizing empire. Furthermore, they "can contemplate and explore, respect and love, an object as substantial as a farm or a native province" (Ransom 1962: 19–20). This reification and striving for terroir links agrarians over the decades, from Bailey to Ransom to Berry to Jackson, to the religious agrarians I met at the food conferences, Koinonia, or Riverview Farm: all are united in love, respect, and care for the local. For the religious agrarians, however, this love and respect is motivated by religious values and grounded within religious worldviews.

CHAPTER FIVE

CONCEPTS OF HEALTH

FROM SOIL TO BODIES, HEALTH MATTERS

Religious agrarians, like ecological agrarians, are concerned about health: of the land, of human bodies, of soil and food, and of society. Furthermore, in their understanding of how these various health concerns are interrelated ecological agrarians influence religious agrarians. For both, the health of one affects the health of all the rest (Wirzba 2003b: x). The concern for a viable and healthy existence provides the focal point for this chapter, where I discuss the aforementioned health concerns. I emphasize what religious agrarians bring to the table that is explicitly different than ecological agrarians: a concept of health that is rooted within religious traditions, for religious agrarians operate within worldviews that contain normative narratives about the health of society. By participating in sustainable agriculture that supports locality, health, and justice, religious agrarians are able to put into practice their religiously motivated values.

HUMAN HEALTH

This section explores various claims about the importance of agrarian thought and practice as these relate to concepts of human health. I divide human health into two categories: physical and spiritual, as ecological agrarians and religious agrarians comment on these two aspects of human health. As I work in these two subcategories, I weave in an analysis of relevant literature from both agrarian types, while highlighting my research showing the strong concern for health that motivates the values and practices of religious agrarians.

SPIRITUAL HEALTH

> "We eat to nourish not only our physical selves, but also our emotional and spiritual selves."
>
> —Savage and Stevenson (2009: 70)

I separate the various rhizomes of health for heuristic reasons, even though agrarians believe they are interrelated. Even within "spiritual" health there is an undercurrent of psychological, emotional, and mental health, which ties into physical health, which ties into social and ecological health, and vice versa. The notion that we are losing the agrarian bond with the Earth is a critique of the dominant industrial culture and its impact on human–land relations and how industrial life has severed a spiritual bond with the Earth. Agrarians believe that a healthy human–Earth bond is expressed in farming and agrarian communal practices and tempers the extreme individualism and rampant consumption that lead to today's social and ecological anomie. Furthermore, they believe that human health, both spiritual and physical, is intimately related to the health of local ecosystems.

Environmental thinkers often assert that human psychological, emotional, and spiritual health is intimately wrapped up with embeddedness in the more-than-human world (Abram 1996; Berry 2006). Some scholars and theorists paint a picture of an imagined golden age when all humans lived in perfect harmony with their local environments and enjoyed abundant health. Others criticize such narratives, labeling them naive and claiming they offer a simplistic and reductionist view of the deep human past. Some scholars point to a need to balance the material/environmental demands of the city with those of the country, instead of fixating on fictitious understandings of human–Earth relations. For example, environmental historian William Cronon argues that instead of pining for an unrealistic eco age, we need to move beyond a dualism between a fallen, faulty humanity and pristine nature and realize we are embedded creatures in a particular place that has shaped and been shaped by human action. This realization should make our lived local environments the purview of our focus in matters of environmental and societal health (Cronon 1995).

Most ecological agrarians would side with Cronon's argument while having sympathies with environmental ethicists and ecopsychologists. Religious agrarians, in contrast, fall all throughout the gamut. For example, Paul Robinson, Ben Peters, and Stephanie Paterson from Koinonia are all significantly influenced by the writings of Daniel Quinn and eco-anarchist Derrick Jensen. These writers argue for a reduction in human numbers and consumption patterns in favor of a bioregional, neo-animist spiritual relation with local landscapes and both have achieved significant followings within the sustainability milieu.

Nonetheless, most religious agrarians would agree with Wes Jackson that "Only by meeting the land's expectations first can society be sustained" (1987: 75). However, once these needs are addressed and the sincere attempt to meet them is begun, questions with no easy answers in today's industrialized world arise about human needs and expectations: what is healthy food? What is a healthy dwelling place? What is a healthy society? How many humans can live in a bioregion, and what are appropriate ways of living that lead to health? Religious agrarians believe that human health contains a spiritual and/or religious dimension, and this dimension includes proper human–nature relations to which the agrarian ideal speaks with intimations for possible answers to these vexing questions.

Such a relation takes on specifically biblical dimensions for Paul Robinson at Koinonia. When asked if he would posit that there are certain ways humans should be living with God's creation, he responded by saying yes. He continued, stating that this view of proper human–God relations affects both his values and practices, as it should for all humans, because "I believe that God created His creation and created nature . . . and in the process told humans [via the Bible] to care for that creation [so we] should obey the commands of God because God seeks to love and that's what he is: He's love." For Robinson the dictate to live out God's love implies that "if you're harming other parts of creation in the process of living, that's not really loving. . . . And I think depending on the interpretation of that really dictates how you look at your relationship to the rest of Creation. If you're assuming that humans are set apart from the rest of it [you're going to treat it differently]" (a similar view is held by Stephanie Peterson, discussed in the last chapter).

Here we see an interpretation of God's commandments in the Hebrew Bible and New Testament, coupled with belief in the creation story of Genesis, that gives humans a clear directive in their role on the planet and how they are to relate to the rest of creation. What is significant in Robinson's view is that God is love, and humans are here to love. Many psychologists and psychotherapists have mentioned the importance of love, acceptance, and nurturance in healthy human development. For Robinson, this love and nurturance comes from God and is to be manifested by humans in their relations with God's creation. Therefore, the spiritual health and well-being of humans is intimately related with how we interact with and treat the rest of God's creation. If we harm that creation, we turn away from God's love, and we are spiritually impoverished.

A Christ-centered view of spiritual health and its implication for proper and healthy human–land relations was expressed by Patrick Smith, who works in the fields at Koinonia, helps milk their cows and goats, and helps slaughter their chickens. He stated, "I take a high view of Christology and see Him as being the Alpha and the Omega, and the reason behind Creation, and it's to me like completely absurd to think about any sort of ministry while neglecting the, the first Gift, [meaning that] if we don't live in harmony with Creation, then there's not much hope for anything else." Thus inner spiritual health is impossible to have without a healthy relationship with creation. This religious value and belief therefore influences Smith's decision to support sustainable farming practices that further the health of his spiritual and physical body, as well as the created body of God.

Similarly, Ruth Jacobs explains how she had recently adopted the Orthodox Jewish practice of ritual hand washing before eating bread:

> To do the ritual wash you pour water over your right hand and then your left hand and you raise your hands up and you bless your hands and then without saying anything else or doing anything else you go to the bread. And then I take the bread I just baked with my own hands and [I] bless the bread. And suddenly there was such a link, to realize that alchemy of bread baking, and that it's drawing from the natural world, and also blessing your hands, blessing what sort of humanity can do, the technology of heat and our hands and craftsmanship. So that also for me was really, really, powerful, to see that Judaism gives us all of this. There's such a meditation in that, such an awareness of just working with whole foods and whole ingredients and creating nourishment for yourself out of it.

Jacobs's ideals of health combine spiritual presence grounded within the Jewish tradition with physical nourishment, seen especially in the practice of her baking bread. Jewish connections between spiritual practice (and health) and physical health were also made at the 2009 Hazon Food Conference, where three different workshops were devoted to the role fasting can play in a healthy Jewish life.

Altogether, the Christian agrarians I worked and lived with used spiritually evocative terms and saw sustainable farming as a task of stewardship and thus religious health more often than the Jewish agrarians with whom I interacted. This is reflected in this section—the majority of the passages I quote about spiritual health are from Christians, because they tended to speak in these terms. This does not mean that agrarian ideals do not factor into Jewish concepts of health, but that physical health was more of a motivator than spiritual health. The religious/spiritual component of Jewish agrarian ideals tended to come out in regard to concepts of justice and in the Jewish injunction to do no harm/do not destroy, rather than in an explicit view of God's commandments and creation and how living in harmony with this was spiritually beneficent.

PHYSICAL HEALTH

Most of the Jewish and Christian agrarians I interacted with used similar language when asked about the benefits of organic farming methods compared with industrial/conventional farming methods and how these respectively relate to human health. Most of these answers also contained considerable overlap with the claims made by at-large supporters for organic agriculture in regard to the perceived physical health benefits organically farmed products have on human bodies. These benefits are seen to be nutritional, neurological, regenerative, and disease-fighting/longevity-inducing.[1]

For agrarians, especially ecological agrarians, the benefits of sustainable agriculture may be spiritual and physical. The physical health component tends to be highlighted, and although the agrarian conception of physical health overlaps with more recent concerns about disease prevention and claims that toxic chemicals are bad for human bodies, the traditional agrarian concern with physical health (where it occurs) is more focused on the ameliorative effects that attend to honest manual farm labor. This labor is seen as healthy for body and soul and is believed to lead to wholesome foods that are equally healthy for the body. Furthermore, such labor is placed within a larger farm ecosystem so that "The enveloping food webs [of a farm and thus a society] will either produce healthy humans or, if we fail to see the connections [between farm, society, and body], diseased humans" (Logsdon 2000: 83).

Many religious agrarians are concerned for their physical health and believe that local, organically grown produce is superior to any variety of conventionally grown produce. This concern is a motivating factor for many CSI-Hazon CSA members. For example, Isaac Segal argued that the value of organic compared to conventional food is that "it's a health factor. Not quite knowing what's going on with the stuff and worrying about the impact on health and more specifically the [health of our] kids." When asked if he is concerned about chemicals, Segal responded, "Yeah, pesticides." When asked where there might be synergies between being a Jew and joining a CSA to get chemical-free organic food, he said,

> Well, probably two things. I mean there's obviously the stewardship of the world, we're supposed to take care of it [see here overlap with some Koinonia members], not trash the place [here there is an assumption that using the chemicals of industrial farming is not healthy for the Earth]. And same things with your body [a shared assumption that industrial farming chemicals are bad for human bodies]. Just like you don't tattoo your body . . . the stuff you put into your body should be healthy as well.

When pressed on his concept of stewardship and where this belief came from, Segal admitted it does not come from the Torah or a rabbinical teaching but is a Jewish duty in the sense of it being a *mitzvoth*.

Another Tuv Ha'Aretz member at CSI, Noah Gottwald, is heavily influenced by the work of Michael Pollan. Pollan's analysis of contemporary food systems inspired Gottwald to search for a CSA, at which point he heard about Hazon-CSI from a Jewish friend who was already a Tuv Ha'Aretz participant. Gottwald shared how he became educated about sustainable agriculture and read more literature about the side effects of industrial agriculture, leading him to believe that organic agriculture should be the default method of growing all our food. As he stated, "Unless you have to, you shouldn't be using chemical fertilizers and chemical additives and antibiotics and all of these things."

Besides being a member of the CSA, Gottwald is on the board of the Georgia Conservancy. He has become very passionate about food and environmental issues. This passion led him to stop eating meat after he learned about the effects of a meat-based diet where the "primary motivator [of this decision] was health. You know, reading about the . . . inhumane treatment of the animals [and how] they're raised and the way that they're kept and the way that they're fed and the way that they're treated with hormones and with drugs was not something I'm interested in putting in my body." Gottwald's experiences show the correlation between educating oneself about the perceived ills of industrial agricultural practices and the desire to switch to supporting sustainable agricultural products, with such desire being influenced by concerns about physical health.

These concerns about health and the perceived benefits of supporting sustainable and agrarian farming practices motivated many of the 600 participants at the Hazon 2009 conference. One woman in the audience during a discussion about sustainable farming practices said, "I'm trying to bring my Judaism into what I do with my hands in the dirt," with the implied belief that healthy farming in healthy dirt will lead to her being a healthy, engaged Jewish food activist. She called this hands-in-the-dirt approach a "new Judaism," and this seemed to be an underlying current of excitement for many religious agrarians with whom I interacted, whether Christian or Jewish: many feel they are part of a new movement within their tradition(s) that are centered explicitly on what I call religious agrarian concerns. Not only are these religious agrarians paying attention, they are actively motivated to put these concerns and the values that underlie them into some sort of practice.

For example, at the 2013 Hazon conference one participant shared with me that she had been exposed to veganism and issues of industrial animal farming, which prompted her to think about the health effects of her diet. While learning about how eggs and dairy were produced, she realized that "physically and spiritually, that's wrong." This epiphany encouraged her to seek out healthier alternatives—physical and spiritual—and this led her to become involved with Hazon. When I met her, she had already been to a previous food conference and was at the 2013 one with her young son and daughter, thus making the 2013 conference a family experience (remember the quote in chapter 1 about the importance of family in shaping food choices). Another 2013 participant echoed these sentiments, when during the panel on "The State of the New Jewish Food Movement," she stood up and shared with the large audience that "we are mistreating life on this planet—human life, animal life, and vegetation." To which one of the rabbis responded, "I think a lot of people don't want to know" about how the planet is treated, especially in regard to food production. As such insights built over the evening's open panel, Nigel Savage explained, "The goal of attaining a sustainable food system is truly a messianic goal." Hazon is laboring toward this goal, and this modern religious agrarian goal represents a new effort in Judaism.

Hazon staff constantly reinforced the idea of "new Judaism" during the 2009 and 2013 conferences. For example, on each day of the 2009 conference (a practice largely followed in 2013) there were three meals served in the cafeteria, always with kosher options, and each day diners found a new "table talk" sign at the table where they sat. These signs shared a Jewish perspective on food or some fact about the benefits of sustainable agriculture. One table talk sign at breakfast dealt with the difference between "natural" compared with "organic" foods. According to the table talk sign, the US Food and Drug Administration regulates natural foods, whereas the US Department of Agriculture regulates organic food and farmers pay for certification. Given the conflicting sources of information and common misunderstandings about what counts as natural versus organic, the table talk sign read: "So what's a conscientious shopper to do? Read the label. How many ingredients are listed? More than five? Are any of them hard to pronounce? Remember the words of Rebbe Pollan: 'If you can't say it, don't eat it'" (quote taken from Pollan 2009). With these daily signs, the conference organizers communicated ideas about what is healthy versus harmful food, grounding this perspective in the work of "Rebbe" Pollan.

These concerns about the relation between farming techniques and human health are growing in the US food culture at large. While statistically still in the minority, more people are buying organic, sustainable food products nationwide and many do so because of health concerns. These consumers are largely convinced that organically grown food is healthier for our bodies than industrially farmed food with its attendant chemicals and fertilizers. Many parents opt for organic foods for their newborns and young children because of concerns about ingesting chemicals. Meanwhile, others are worried that some industrially farmed food items can be carcinogenic due to the large amounts of pesticides and chemicals used in their production.

Many religious agrarians hold such concerns, including members of CSI's Tuv Ha'Aretz and Dorothy Goldstein, has heard about most of these concerns directly. Goldstein was instrumental in bringing the Hazon partnership to CSI and organizes the CSA pickup at the synagogue. She is in charge of signing people up for the CSA. In her experience as the Tuv Ha'Aretz organizer, "I think concerns about health and toxins [are] huge to a lot of people . . . And a lot of people come to this that you really wouldn't think of because they've done the homework and they realize that they're pregnant or they're nursing or they have young kids or they have some other health issue and they want to go organic. . . . A combination of organic and local and small business supporting, I think it's that sweet spot of all of those components that's very important to people and very important to our membership." Goldstein explains that many Tuv Ha'Aretz members at CSI are motivated to join for personal health reasons. Yet significantly, she points out that the members are also from all walks of life, are typically informed about health and food issues, and combine this concern with supporting small local farmers who they believe grow healthy food in healthy ways.

FARM HEALTH

Agrarian thought is intimately related to concepts of holistic farm health. These concepts enjoy a long standing in sustainable agriculture circles and networks, with an underlying tenet being that the farm is an ecosystem that needs to be properly managed and cared for if it (and the soil, animal, and plant life dependent on it) is to be healthy. We have already seen evidence of this narrative in the Grow and Behold pamphlet from the last chapter that provides a religious agrarian understanding of farm health. Metrics of ecological agrarian health for a farm include clean water devoid of chemical and manure run-off; healthy and vibrant microbial life in the soil; seasonal and polycultural plant and animal breeding and planting; minimal loss of plant and animal life to pests, diseases, and parasites; a sense of fulfillment and challenge for those farming the land; and healthy products offered to the wider community to consume.

Many books, articles, and movies speak to these ideals and to the hard work that is required to meet them. Significant parts of this volume have cited much of this literature, and the prior chapter on locality equally explored this issue while highlighting the practices undertaken to create farm health. Therefore, I make only brief mention of concerns about farm health (and the values that underlie them) before moving on to concerns about the health of society. I do so by using the work of Maria Rodale, whose grandfather is J. I. Rodale, the founder of one of the magazines that helped develop consciousness about organic foods in the United States: *Organic Farming and Gardening.* Maria Rodale has followed in her grandfather's footsteps and in 2010 released her version of *Organic Manifesto.* This book begins with a common narrative about the health benefits of organic versus industrial agriculture:

> The technological mindset that would dump billions of pounds of deadly chemicals onto the soil, and mix the genetic material of different species, and build factory farms where livestock are treated like industrial commodities, and clone animals in order to give them a uniform size, has a deeply arrogant view of the natural world. It regards Nature as something to be conquered and controlled for a short-term profit. The organic movement embodies a different mindset. It takes the long view. It seeks the kind of profit that can last for generations. It regards the natural world with a profound reverence and humility. It aims to work with Nature—and considers the whole notion of controlling Nature to be absurd. At the heart of the organic movement is a belief in the interconnectedness of things. (2010: xi)

While she neglects to comment about the perceived salubrious or non-salubrious benefits of large-scale organic agriculture, Rodale nonetheless mentions the underlying values and worldviews behind these two types of agriculture. Her narrative is similar to the agrarian narratives—religious and ecological—of those I met in my research, especially in regard to the concept of health.

For example, Rabbi Elijah Greenberg of B'nai Israel in Gainesville, Florida, explained that in the Torah is found "the commandment to let the crops lay fallow every seventh year. That's a religious commandment, a religious imperative, the results of which we understand was the next year crops would come back stronger, and we now understand the science behind that. But the religious directive exists right there in Genesis and is repeated in Exodus and Deuteronomy, to have respect for the land. And to let the land have a Sabbath just as we human beings need a Sabbath." This ancient practice indeed has its modern counterpart in sustainable farming practices, seen when modern farmers let fields lie fallow and rotate their crops (a practice followed by some industrial farmers), and is now being advocated for in Hazon's Shmita program. This farming practice leads to increased health of microbial life in the soil and thwarts certain blights and root-based diseases that typically develop if the same crops are planted in the same fields year after year.

Rodale explains her concerns about health in part one of her book, titled "The Great Chemical Experiment." She writes that, "We are all in the same situation to varying degrees. We are all being poisoned, contaminated, sterilized, and eventually exterminated by the synthetic chemicals we have used for the last 100 years to grow our food and maintain our lawns, to make our lives easier and 'cleaner' and our food 'cheaper'" (Rodale 2010: 4–5). She takes this concern for physical health and links it to farm health, showing that within agrarian circles the concerns for and beliefs about health regarding agricultural issues are interrelated.

This concern is echoed by Beatrice Tischler, who shared with me her belief that "We're poisoning the land. I lived in Wisconsin for eight years in an agricultural state and you really do, there were times when the rivers and lakes were closed because of nitrate run-off from all the cattle farms." Thus, Tischler's lived experience with what she perceived as ills of industrial farming have mixed with her Jewish identity and motivate her to support sustainable agriculture. She also explained that this same identity follows the concept of *tikkun olam* ("to heal the world"). For Tischler, this concept means that Jews "have an obligation—we're also an agricultural people, the Jews—we have a strong obligation to treat the land, to rotate crops and not to deplete it and not to poison it." Ascribing to this Jewish teaching suggests that this Jewish religious agrarian would be sympathetic to Rodale's manifesto.

Rodale continues by arguing for the abandonment of chemicals and adoption of organic growing methods. She provides evidence that once organic farming methods are adopted and soil is restored, crop yields become comparable to the latest chemical yields and are actually higher in years of drought and flood. She also points out that organic fields require only 30 percent of the fossil fuels used in conventional fields when growing corn and soybeans; organic fields need 15 percent more labor inputs (which supports local economies and thus societal health); the net economic return is equal or higher; and last, organically farmed fields store more carbon and nitrogen in soil that can help offset global warming (Rodale 2010: 151–22).[2] She uses these comparative statistics in grounding her manifesto and feels they offer a compelling picture of farm, individual, and societal health.

Such a comprehensive view of the importance of healthy human–land relations is shared by Lisa Jones, executive director of Koinonia, who states that "if you're caring for the Earth, you're in partnership with the Earth, you're not battling the Earth. And it just seemed to me, to pour things toxic on the Earth, you're battling it, [you're saying] 'I'm going to dominate you and you're going to produce more' . . . Just knowing that what they did first at Koinonia is that they came and bought a broken down old farm and nurtured it back to the health . . . made a lot of sense to me." For many members of the community, this history is a guiding inspiration for the health-based religious agrarian practices of Koinonia today. I witnessed another expression of Jones's desire for the healing of Koinonia when she introduced herself to the participants of the permaculture training I attended in 2010. When welcoming them, Jones said, "You are part of the healing process. The physical and spiritual healing of Koinonia." This means that for her, participation in and sharing of the religious agrarian practice of permaculture is leading to the healing of the land and spirit of the entire community: human and more-than-human. We now turn to concepts of societal health from agrarian perspectives, seeing that ecological and religious agrarians speak much of the same language in respect to this metric of health.

SOCIETAL HEALTH

Agrarian concerns about societal health include concerns about the health of various types of representative democracy, the ills of industrialization of farm and workforce, opprobrium toward overconsumption and lack of concern for and identity with a local place, and lamentation over the loss of societal respect for good farmers who steward a part of the community. Given these concerns, "civic agriculture . . . is associated with a relocalizing of production. From the civic perspective, agriculture and food endeavors are seen as engines of local economic development and are integrally related to the social and cultural fabric of the community. Fundamentally, civic agriculture represents a broad-based movement to democratize the agriculture and food system" (Lyson 2007: 19). Thus, concerns for relocalization (versus industrialization) are by default concerns for democracy, sustainable and local economic development, and community health. Agrarians believe that their vocation is part of a larger panacea to bring these social health concerns together, midwifing society at large from its fixation on individualism and productionism to justice and a healthy citizenry.

At Koinonia, Katie Taylor passionately addressed these concerns during our interview. She said, "the way that America thinks about food is so wrong. . . . It's instant in gratification. And it's a right that you can only get when you work." Taylor argues that compared to this "me-first," unhealthy relationship to food, "I feel like permaculture and organic gardening is going back to part of who we are: we have to take care of [the Earth] just like we have to take care of ourselves, we have to be gentle, we have to be stewards." To be a steward means, "to take only what you need. To not overwork the soil and not to overwork the earth is not to destroy things by taking so much to make a profit."

Here Taylor is concerned that many in the United States see food as a privilege to be enjoyed if they have the money to buy any food they want, and furthermore, the food they buy tends to be unhealthy for soil, body, and society. Because of the productionist paradigm of industrial farming where commodities are grown only for profit while damaging the Earth and soil, Taylor is concerned that human communities are also being damaged, with the poorest of the poor suffering (see connections to agrarian concerns for justice and locality). These values and beliefs, based on a larger Christ-centered values of fellowship and justice and stewardship, inspire Taylor's support for religious agrarian practices like permaculture.

Nigel Savage expressed similar sentiments from a Jewish perspective during his opening reception speech at the 2009 Hazon Food Conference. At this reception he told the audience, "As a people we have ancient gifts to offer the world—without arrogance—to make a sustainable world. As families and individuals we have roles to be a blessing to the world." Here Savage underscores the link of sustainability and health by focusing on the individual, family, and community.

In Savage's vision, the emerging Jewish food movement is part of a larger societal movement toward sustainability, where food functions as the vehicle of transformation. With their willingness to work with and include non-Jews, Hazon is forging lived networks and helping shape the sustainable food movement from a distinctly Jewish platform. Hazon is therefore part of "the community food system movement [which] is challenged by the need to bring together many distinctly different local food system efforts into a definable community food mosaic. This mosaic illustrates the vitality of what is emerging as a social movement as well as a fledgling political agenda" (Maretzki and Tuckermanty 2007: 333). The Christian and Jewish agrarians I met are indeed part of a social movement that has political and societal prescriptions, and they actively put their religious values into practice in these two realms.

Wendell Berry expresses similar thoughts about the health of society from an ecological agrarian perspective. In his words, "the best advice may have been given by Hippocrates: 'As to diseases make a habit of two things—to help, or at least, to do no harm' . . . To help, or to try to help, requires only knowledge; one needs to know promising remedies and how to apply them. But to do no harm involves a whole culture, and a culture very different from industrialism. It involves, at the minimum, compassion and humility and caution" (2005: 65). Religious agrarians are helping build this nonindustrial culture as well. This is evident in their lived networks, lifestyle choices, and interpretations of their traditions. Through these vehicles, religious agrarians attempt to act with compassion, humility, and caution. They also have strident critiques of the industrial mindset and its manifestation in policy and farming.

One area of concern for health where this expresses itself for many religious agrarians is their sympathy to and support of the critique of concentrated animal feeding operation (CAFO) farming and industrial slaughter of animals raised in these environments. They support the work of Eric Schlosser (2005) and his research about those who work on slaughter lines and how these workers (typically immigrants) are emotionally affected by their work and how they are shaped by this workplace trauma that then affects their families and society.[3] They are also typically in agreement with and have been influenced by the work of Michael Pollan and his criticisms of the US Farm Bill (criticisms shared by Wendell Berry and Wes Jackson [2009]) and CAFO farming methods (Pollan 2007).

Furthermore, some religious agrarians are part of lived networks with Slow Food International. This membership helps support the latter's goal to create healthy societies through the enjoyment of communal meals made with locally sourced, sustainable food eaten at a leisurely pace. A workshop at the 2009 Hazon Food Conference highlighted this link by investigating the Jewish concept of hospitality (*hachnasat orchim*). Koinonia has been living out the ideals of slow food in their own Clarence Jordan–inspired way since the beginning of the community and to this day practice hospitality around shared meals. By supporting farmers' markets and joining CSAs, many religious agrarians are engaging in lived practices that from their perspective directly contribute to societal health.

According to one of the leading agrarian writers, Gene Logsdon, the contributions that entail to a modern agrarian lifestyle are needed if we are to approach sustainability as a society. In fact, he is so concerned about societal health that he claims, "Sustainable farms are to today's headlong rush toward global destruction what the monasteries were to the Dark Ages: places to preserve human skills and crafts until some semblance of common sense and common purpose returns to the public mind" (Logsdon 2000: xii). By supporting and participating in the agrarian ideals manifested by sustainable agriculture, the religious agrarians I research are contributing common sense and common purpose to society while realizing they live in agricultural networks that are "context[s] of multiple obligations and needs" (Sanford 2011: 300).

ANIMAL AND SOIL HEALTH

Many religious agrarians are sympathetic with criticisms of CAFO feedlots and the way industrial farm animals are raised and slaughtered. Koinonia is actively breeding and raising its own animal herds and flocks, which they attempt to humanely slaughter, dress, and consume on site. Their animals are free range, grass fed, naturally cared for with minimal use of antibiotics (only when their health is at stake and all attempts to heal them naturally have been used), and receive human companionship throughout the day. Riverview Farms treats their hogs and cows the same way. The Southern religious agrarians I interacted with who eat meat all claim to buy free-range, organic, antibiotic- and nitrate-free meats and eggs (the Jews I interviewed who follow kosher guidelines eat no pork products).

Many religious agrarians are also concerned about how industrial farming treats soils, as seen in some of the previous quotes about "poisoning the land." For example, when asked about the perceived benefits and values of organic farming methods, Noah Gottwald responds, "I guess the legal definition of organic and the holistic understanding of organic aren't necessarily always one and the same thing, but treating the Earth right, whether that's for plant farming and harvesting crops and not, and not having too much pesticides and run-off, or with animal farming and treating waste properly and treating the animals properly and not giving them hormones to make them grow faster and not giving them antibiotics so that you can keep more of them in a closer knit area without them getting sick." Gottwald's answer is influenced by his own experience of driving past hog farms in North Carolina, where "you don't have to know about the downstream impacts on the environment to know that that smell is unhealthy and unnatural and can't be good for the people living around it. . . . And it can't be good for the water supply that's getting, getting washed through with it. And it can't be good for the wildlife that are around it. I mean, everything about it can't be good. The only thing that it's good for is the pocketbook."

Gottwald's response evidences concern for the land, animals, and the perceived impacts that industrial farming has on local wildlife and waterways. It also contains a criticism of the "productionist" model of animal farming. What is significant about this answer is that he exhibits knowledge about the contested meanings of organics and his perspective on the issue has been shaped by lived experience.

Before leaving Koinonia, Ben Peters was on the garden crew and spent his days working in the organic garden and permaculture-designed fields of the campus. He worked closely with Joel Taylor in deciding what would be planted, where, and how much, as well as deciding what animals to range where on any given day. His research into industrial versus organic farming, coupled with his experience at Koinonia, led him to tell me about the value of organic agriculture: "[At its] best [it] is part of a healthy human ecosystem, [and] its main function would be to provide food for humans. Conventional agriculture would be part of the unhealthy system of earth exploitation and its main [function] would be to produce money through the facilitated destruction of earth's resources. And also to make food." For Peters, conventional agriculture is largely about profit motive for corporations, where this fixation on money engenders unhealthy farming practices and technologies like "fertilizers or in plowing, [because] excessive plowing might produce short-term increases in yield but it can lead to the desertification of an area." Peters feels that large-scale organic practices can have the same results, so scale is an issue. He also recognizes that, "some conventional farmers really care about the earth and some organic farmers are doing it just for the money."

Peters's views have been shaped by reading literature about sustainable and industrial farming and by his experience living at Koinonia and seeing "the pecan trees, and knowing it [is] dangerous to walk after some of the stuff they did. Like with the spraying and stuff." He is referring to yellow signs erected under the pecan groves after the trees have been sprayed with industrial chemicals. The signs warn people not walk in the fields for twenty-four hours to lower their chances of getting sick from the chemicals. This lived experience, coupled with seeing industrially farmed fields and sustainably farmed fields next to one another, has influenced Peters's views about soil health.

Another member of the farm crew at Koinonia, Patrick Smith, earned a degree in animal science and worked on a large-scale dairy farm in the Northeast. Since moving to Koinonia with his family, Smith has been heavily influenced by Jim Everett and enjoys working with the animals and in the fields. When asked about the value of organic farming methods, Smith opined that it is "the sustainability factor, the health factor, eating foods that are more nutritious, that don't have the negative aspects of chemicals being ingested into the body and whatever kind of unforeseen damages they do to the human body. We all feel safer eating foods that are naturally grown without chemicals." Of central importance to my argument is that Smith is an exemplar of a religious agrarian worldview, in that these views about health are directly embedded in a Christian perspective. As he explains, "I think [health] lines up with a Christian view of stewardship of creation, caring for the soil, supplementing the soil."

Like Peters, Smith's views have been shaped by living at and working in the fields of Koinonia. According to his analysis, there is a clear difference in the health of soil and food between the community's sustainably farmed fields and those fields that were in industrial rotation, which are mainly the pecan orchards. These beliefs and observations became a typical topic of conversation amongst those working in the fields. During these work-based discussions many members expressed excitement at seeing the community embracing more permaculture practices and at seeing the direct results these practices were having on the health of the community's soils and crops.

Contemporary soil science can be interpreted to support the religious agrarian concern for soil health. For example, Rodale writes, "What we call 'soil' is a living thing. Just one tablespoon of soil can contain up to 10 billion microbes" (2010: 11). She adds, "Right now, soil scientists understand less than 1 percent of all the living things in the soil" (ibid.: 11). In her analysis,

> Agricultural chemicals have statistically and significantly been implicated in causing all sorts of cancers, behavioral problems, attention-deficit/hyperactivity disorder, autism, Parkinson's disease, reduced intelligence, infertility, miscarriage, diabetes, infant deformities, and low birth weight. And with endocrine disruptions come

> genital deformities, early puberty, gender "issues," and, again, diabetes and cancer. But all this research comes from the few scientists courageous enough to swim against the tide, to resist the easy funding offered by chemical and pharmaceutical companies and the pressure of their peers who rely on that funding. (ibid.: 36–37)

Therefore, for supporters of sustainable agriculture, it makes sense to put into place farming practices that honor and support life in the soil; thus, adding toxic chemicals to the soil depletes this microbial life. Daron "Farmer D" Joffe, with his syncretic mix of Jewish and biodynamic beliefs, expresses such views about stewardship and concern for soil health:

> But down to [the] basics it's about being good stewards of the planet. It's about our obligation to till and to tend with responsibility about generations who are going to follow. Putting chemicals into the food and soil and water, it does not make good sense. I don't care what your political [beliefs are] or where you stand on that side of the fence, it's not caring for creation. It's a responsibility we have as humans—it's about as basic as it gets. And you want to take care of yourself, your family and children, raise them healthy and clean. So I guess I have a belief that healthy clean food can't really be [grown by] conventional, industrial farming techniques. And it also makes sense from an ecological standpoint, and as far as a business, you invest in your soil. You invest in the biology of your soil, and science and soil biology and even the golf course industry now accepts that the biology of the soil is the most important aspect of weed, pest, and disease management. It all starts with the quality of your plants based on the quality of your soils. The quality of the humans and animal health is based on the quality of the soil.

Joffe's answer provides a snapshot of the agrarian concern for health and the interconnectedness of soil, animal, food, society, and ecosystem. He passionately points out that it is a moral imperative, for present and future generations, to properly steward the Earth. His views are very similar to those of Ben Peters and Patrick Smith—all three operate within a religious agrarian worldview that is influenced and shaped by their experience living and working on sustainable farms.

Many of the talks at Hazon's 2009 Food Conference shared this interrelated concern with soil, animal, and social health. However, one topic in regard to animal health that dominated the proceedings was the status of the kosher diet. Two discussions in particular stood out in this debate, and these were the fidelity of kosher processing plants to kosher law and the debate about a sustainable kosher option.

Those Jews in the United States who keep kosher were rocked when it came to light that one of the nation's largest kosher processing plants, Agriprocessors in Postville, Iowa, was shut down on May 12, 2008, because of Immigration and Customs Enforcement violations. Agriprocessors was employing "illegal alien" workers at the plant, so its status as being a trustworthy source of kosher products (in a ritually slaughtered and social justice sense) was called into serious question. Some Conservative Jewish groups have responded to this event by developing their own kosher certification that is based on Heksher Tzedek, a certificate of social justice (the new seal is Magen Tzedek; see next chapter on justice) in how workers and *shochets* are treated and justice in how animals are slaughtered.[4]

To be kosher certified, a kosher inspector called a *mashgiach* must inspect a plant's equipment and processing methods. Furthermore, a shochet, or ritual slaughterer (who must be Orthodox for food to be kosher) must kill each animal along various kosher laws for the animal products to be considered kosher. Most Conservatives and all Orthodox Jews in the United States follow kosher dietary laws, so the Agriprocessors event was a huge issue for US Jews. In short, the evidence that emerged suggested that one of the largest suppliers of kosher products in the nation was possibly not following kosher law and was definitely not following legal hiring practices.

So while "Kashrut is a constantly evolving construct" (Fishkoff 2010: 10–11), it is evolving at a more rapid pace due to the Agriprocessors event and the development of the contemporary Jewish food movement. Arthur Waskow popularized the concept of "eco-kosher" in the 1970s, and the concern for animal products raised in ecologically conscious ways is meeting with the Jewish farmers, foodies, and activists of Hazon so that a new investigation of what kosher means is happening. This is seen in the work of Gavin Sanders, whom I met at the 2013 Hazon conference. Although there is a long tradition of vegetarianism in Judaism, there is also a tradition of kosher slaughter. This tension exists in Hazon, and for those I met who do eat meat, most search out organic meat where and when possible.[5] They also have to find kosher meat, which makes it a harder task, so the demand for sustainably raised meat that is kosher certified is on the rise within quite a few Jewish communities, and meeting this demand is one reason Anna Hanau and her husband began Grow and Behold.

For example, Rabbi Greenberg notes, "We've had relationships with a kosher butcher for many years, and the closest is in Atlanta. And we have a relationship where he brings kosher meat to this community every six weeks, and we made a commitment as a community to support him. And for those people who observe *kashrut*, he's our source of kosher meat." This relationship with a specific shochet made it so the synagogue found it easy to embrace a similar CSA relationship with a farmer in the local community. As Greenberg explained, "So we've already had these kinds of relationships in the past. This is nothing new for us."

Meanwhile, Naomi Eisenberg explained that she stopped buying kosher meat because for her kashrut is not only about cleanliness and keeping people healthy but also about cruelty to animals and to make sure they suffer as little as possible. As she explains, "If they're going to be bred to keep us fed and healthy, then they shouldn't be penalized by that to keep us fed." PETA helped expose Agriprocessors in 2004 via footage from undercover cameras so that when the story finally broke in the news, Eisenberg stopped buying kosher meats except for chicken. Today she buys free-range, antibiotic-free, hormone-free kosher chicken from Whole Foods. Her evolution in regard to her relationship to kosher meats evidences a familiarity with her tradition and suggests that her observance of kosher laws inspires her to find sustainably raised and humanely slaughtered products. In fact, one of her goals is to find a shochet for the meat raised on Riverview Farms so that this meat can become a part of the CSA offering.

Ruth Jacobs expresses similar distaste about kosher products that are mass produced and artificially flavored. For her, "I'm endlessly disgusted and grossed out when I see what's kosher so often . . . it can be so incredibly processed, and so much just junk, just shortenings and other things to create things that otherwise wouldn't be acceptable." This lived experience motivates her to partner with Hazon and advocate for healthy kosher products so that "if we can move this community away from the processed kosher foods and into whole foods and say those are really actually more kosher," this would be a positive way to encourage healthier Jewish diets.

Jacobs sees a reinterpretation of kosher to be consistent with the deep past of the Jewish tradition. As she explains, "You know, the Jewish cycle is so intricately linked to agriculture and food, and so when we live in a world where things are processed, and we can in some ways defy maybe the rules through food technology, I think that our bodies suffer in terms of health, and I think that spiritually something is lost." Given its agricultural past, for Jewish agrarians the Jewish tradition serves as a vehicle to embrace agrarian concerns about health where this embrace can have spiritually beneficent results. In Jacobs's approach, we see links between personal, societal, and ecological health, coupled with a desire to redefine what kosher means to include locally raised whole foods. From her position at B'nai Israel, she is able to put her own religious agrarian values and convictions, her lived experience, and her Jewish identity into practice while influencing the lifestyle choices and health of others in her community.

The most damning critique of meat eating and animal agriculture and the impact animal agriculture has on the environment I encountered came from Rabbi Joshua Hillel of CSI. He passionately explained, "Kosher meat is, well it's disgusting, it's gross, it's Postville, Iowa, it's large-scale factory industrial production. It's nothing short of cruel torture of animals their whole life." In his eyes, the shochets who work at these large-scale industrial kosher plants are "not the pious old saintly Jews. These are the money-grubbing mafia kind of Jews who will kill and kill and kill and they make their living in a disgusting horrible way built on the backs of the torture of God's innocent creatures. So to pretend like they're the saintly old rabbis of the Talmud, I mean give me a break! That's not what it is." Because of the distancing of animal to plate, Hillel recognizes that there is "a disconnect between the deepest Jewish values and the way that we actually live our Jewish life." He also recognizes that industrial agriculture, especially kosher meat production, is "a desecration of our values! [As Jews, we should not] change [our] values to accommodate the socio-economic realities of meat production in the twenty-first century." In his words, Jews should "stop eating meat because it's contrary to [our] values! You stop eating that way and you say, 'even though our ancestors didn't know this, and wouldn't have said it this way, our tradition compels us, mitzvoth, obligations, the Torah compels you to be sustainable in [our] agriculture.'" Rabbi Hillel explains that the Torah

> compels [Jews] to take care of farmers and to not put small farms out of business . . . Who said you're guaranteed cheap meat and unsustainably cheap food? Where is that written in the Torah? Where is that guarantee? All it says is "You cannot be cruel to animals." You're being cruel to this animal. I don't care if that means the meat is going to cost you a hundred times more and I don't care if it means you never get to eat meat again. Those are the ethical and religious imperatives, *not* the availability of cheap kosher meat. That's not an imperative.

This quote is important to my argument for several reasons. First, it shows Rabbi Hillel's concern for animal health while also showing a modern challenge to the perceived corruption of what kosher has become under the dictates of industrial agriculture. It also shows once again that many religious agrarians perceive the issues of health, locality, and justice to be intertwined. In this case, there is concern for the health of animals and the concern for justice in supporting local farmers. Last, this passage directly points to the issue of values: for Rabbi Hillel, Jewish values have been compromised by Jewish support of industrial agriculture and its for-profit model. Such compromise goes against the teachings of the Torah and the commandments of God, which "compel" Jews to be "sustainable in your agriculture." To keep kosher in today's industrialized food world means that not eating meat is the only possible way to follow the spirit of kosher: only by not eating meat can a Jew follow the commandment to not be cruel to animals.

I want to conclude this section on soil and animal health by relating the compelling story of Aitan Mizrahi, a young farmer attempting to be sustainable in his work due to his Jewish values. Like Daron Joffe and other young farmers at the Hazon Food Conference, Mizrahi is a former fellow of Adamah. He raises goats, and "Like most of his friends in the new Jewish food movement, he is intrigued by traditional Jewish rituals but observes only those he finds personally meaningful" (Fishkoff 2010: 300). In Mizrahi's words,

> I very much identify as a biblical Jew. The beard is symbolic of my Judaism; it reminds me of who my ancestors were and how they would walk the hills of Judea with their goats and their sheep. They had a deeper relationship to the land, and how that land connected them to the holy spirit of God. I very much feel that my work connects me to HaShem ["the Name"—how some Jews refer to God in casual conversation]. I know how I want to raise my animals, and I believe that how the animal is raised transfers to its meat and milk. If the animal is nervous or stressed, her milk will taste poorly. But when my goats are out in a pasture and are getting good hay and clean water, you can taste that the milk is a lot fresher, a lot cleaner. And I think that translates to the meat as well, on a spiritual level. (ibid.: 300–301)

Such sentiments about the way animals are treated and how this translates into a superior-tasting product that is healthier are shared throughout food activist networks and circles and increasingly within religious agrarian networks. Although involved in religious agrarianism in different ways, from farmers attending a Hazon Food Conference to participating in a permaculture training at Koinonia, all believe that how we farm influences the health of the body, the spirit, the community, the soil, the animal, and the land. They believe that there are clear religious-based values that require people farm in healthy, sustainable agrarian ways.

PLANETARY HEALTH

The last health concern of religious agrarians is that of the planet. This includes the health of the global climate and of the health of regional and global rivers and supplies of groundwater, for agriculture has a strong influence on climate and water. For example, up to 60 percent of all the freshwater in the United States is used for agricultural purposes. Such demand is precipitating a rapid drawdown of the nonrenewable Ogallala aquifer. Furthermore, many of the industrial chemicals used on farms end up leaching into groundwater and thus the water table in the form of nitrate and other chemical run-off.[6] Sometimes the run-off contains animal waste and excrement or chemicals that stew in manure lagoons on factory farms, which can also leach into groundwater tables. Studies suggest that agricultural chemicals account for two-thirds of all water pollution. High concentrations of chemicals lead to hypoxia, so that marine dead zones have doubled in size every ten years since 1960. This doubling correlates with the increased use of ever-more toxic chemicals by industrial farmers. On top of this, intensive animal agriculture is a larger contributor to global warming than is transportation, meaning what we farm and how we farm it truly does have an impact on a planetary scale that future generations will have to deal with.[7] Once again we see that with food choices, everything is at stake.

Given these planetary effects, it is no surprise that a nascent yet articulate lived religious agrarian response is developing. For example, Paul Robinson says, "when I walk into any situation the first things that pop into my mind . . . the first filter I look at the world through is a Christian and God-centered view point so I'll notice things that don't really fit into obeying God's commands and His instructions on how to live. And then the next thing that pops into my head is how is this damaging the earth and the environment? So immediately when I walk into a room I notice like lights are on, appliances are on, and possibly could be turned off." Robinson's clearly inspired Christian and theocentric values lead him to make lifestyle choices and practices that are in accordance to living within God's commands and caring for God's creation. This search eventually led him to live at Koinonia, where he can more closely follow a sustainable lifestyle of fellowship and live out God's green commandments.

Another person I interviewed is Russell Bain, a self-described Christian mystic whom I met during Koinonia's permaculture training. This was Bain's second training at Koinonia, and part of what brought him back was the Christian atmosphere. When asked how being a Christian and following the example of Jesus translates into starting his permaculture business and having passion for the Earth, Bain answered,

> Because I believe that Jesus was connected in a perfect way with nature, with everything and with God, [so] I believe that when we connect with Earth in a perfect way, we can live in a perfect environment. If we allow ourselves to be natural, which would be removing our fears and anxieties, and be connected to nature, we'll be closer to nature. I think the only way we can really be one with nature is to allow nature to be nature. And whenever we are disrupting it, whenever we're not allowing it to be natural, then it disturbs the entire system. Not just the ecosystem, but our entire permaculture system: Earth care, people care, fair share, the whole thing. It's why Earth is so important. It's part of everything, it's all connected, and if we want to be closer to God, then we have to protect the Earth as well.

This viewpoint was expressed by one of the panelists from the vegetable monologues, who explained that God's miracle "exists in the simple presence of plants and fruits and humans on this planet."

A similar ecological hermeneutic was present at Hazon's 2009 conference during a Shabbat prayer service led by Rabbi Steve Greenberg. He explained that Shabbat is the only commandment Jews have to do and is punishable by death. According to him, observing Shabbat is about facilitating the six-day-a-week movement of "doing" to a divinely mandated day of simply "being." In this rabbi's analysis, YVWH, or Adonai, means "is-was-will be = being." He interprets the Sh'ma (Judaism's most popular prayer that reaffirms their monotheistic belief in Yahweh) to mean "love" and argues it is a "profoundly ecological" prayer. For Greenberg Jews believe that God is Being, and according to the Sh'ma, Being is One; therefore, all things are of one creator and are interconnected.

The rabbi encouraged us to ponder this deeper meaning of Shabbat and the Sh'ma. "We all love our own being. Can we love all other beings the same way?" This question brought in an ethical component that challenged us to look at our relations from this interconnected perspective of God's love for all of his creation. In Rabbi Greenberg's theological understanding, the Sh'ma becomes an occasion for Jews to reflect on God's oneness, which is manifested in the interconnectedness of creation. In this theocentric understanding, there is an ethical demand to extend the love we feel for ourselves to the rest of the created world so that living in ways that contribute to the health of the world becomes a vehicle for recognizing the oneness of God.

Moreover, for Greenberg, the Sh'ma prayer, and the Sabbath as reminder, tells Jews that they must love God and do so all the time, so much that they are to bind it to their bodies and doors and teach this love to their kids. He pointed out that in Deuteronomy 6:5–9 and Deuteronomy 11:13–21 it is written that God loves His chosen people and gives Israelites abundance but that the Israelites must be faithful—if they were to stray by worshipping others (or in a modern context by worshipping money or watching too much TV), then rain will not fall and crops will not grow. Thus, God's love and the recognition of God as Being is an ecological teaching that should be remembered (re-member: to re-join with God) every Shabbat. I found this to be a very explicit teaching about the planetary and ecological implications of a Jewish view about God such that it was obvious that this theological perspective motivates Greenberg's work with Hazon.

Similar views about interconnectedness are legion within agrarian writing. For example, Gary Holthaus shares that, "Soil, water, air, and light. These are the things without which neither agriculture nor a society of any kind can begin or continue. Talk about self-interest! Our care for them is the outward and visible sign of our care for ourselves, and one indicator that we have a future on this planet" (2009: 6–7). Furthermore, "Farmers are a species in an environmental, economic, social, psychological, and spiritual ecosystem that includes all the species from soil microorganisms, creatures small and large, and plants to neighbors, nearby towns, institutions like schools and churches and merchants, and large urban areas, as well as wilderness, geological formations, and light from distant stars. The water we depend on for agriculture and sustenance is related to it all, too, never quite contained, even in the quiet sloughs and backwaters of our major streams" (ibid.: 174–75). Thus, "a cosmos where mono-anything is doomed to failure" (ibid.: 176).

Most religious agrarians share this expansive view of health, from the microbial to the personal to the societal to the planetary. Their understanding of health tends to be nuanced and is built on lived experience, personal research, and religious values and teachings. They also believe that health is threatened, even destroyed, by industrial society and agriculture. This "civilization," as Ben Peters argues, "is killing the planet, so it's killing us." The problem with industrial culture, he adds, is that it is not based on "natural patterns." In contrast to the destructive practices and values of industrial agriculture, Peters points to Jesus's example of harmony with creation as being a model for sustainable agriculture.

Rabbi Greenberg offers a similar analysis from a Jewish perspective, stating, "I think buying locally, feeling a connection with the land, growing locally, recycling, treating the environment with care [is what] I understand us to be commanded to do." Lisa Jones believes "we as humans are called to be in partnership with the Creator to re-create the garden. And we have done the opposite. . . . I think we're on the brink of either making a decision to go ahead and be a partner and re-create the garden or go ahead and finish the, the job and kill this planet." As executive director of Koinonia, Jones is actively working with the rest of the community to live as sustainably as possible, as such a religious agrarian lifestyle helps re-create the garden and partner with God.

Whether inspired by the commandments of the Torah and the rabbinic tradition or from Protestant readings of the Bible, religious agrarians are committed to bringing about the health of soils, animals, human spiritual and physical bodies, communities, and planetary ecologies. They are equally committed to issues of justice, the subject of the next chapter.

CHAPTER SIX

JUSTICE FOR ALL

FROM SOIL TO WORKER, FROM INDIVIDUAL TO COMMUNITY

> Agriculture is in the foundation of the political, economic, and social structure. . . . If agriculture cannot be democratic, then there is no democracy.
>
> —Liberty Hyde Bailey (2009: 103)

A CONCERN FOR JUSTICE

In this chapter I examine the importance of justice in religious agrarian values and habits of thought and the ways that agrarian practices and lifestyles embody concerns for equality, fairness, and empathy toward those most vulnerable to abuses of power. Such concerns mobilize many sustainable food activists and provides motivation for many of their practices. This holds for ecological agrarians, and even more so for religious agrarians. For the latter, an emphasis on justice stems from identity formation brought about by being part of religious traditions that have long-standing critiques of the manipulation of power, greed, hubris, and acts of injustice. Many of the people I interviewed have read religious scripture, such as the Hebrew prophets, where critiques of power are central. Many have been raised by religious parents or attended and participated in religious services where the value of treating other humans in just ways was emphasized. Many desire to extend such compassionate and ethical care toward nonhuman life forms and toward what is perceived to be a sacred, holy, and/or divine creation. By participating in sustainable agriculture, religious agrarians are able to put the values of religiously informed justice into embodied practice.

POLITICS AND FOOD

Many agrarians believe that "Food is political," because "To change the food system requires organizing" (Lappé 2010: 242). A clear commitment to social justice motivates many sustainable food activists to engage in political actions and organizing in addition to their farming activities. The simple act of creating self-regulated certifying agencies in the early days of the nascent organic movement was itself a political statement, as well as being a critique of the perceived injustices of the industrial food paradigm and its attendant practices. Such critiques and political mobilizing have been present in sustainable food circles for decades, and these concerns are still front and center, as witnessed in the recent March against Monsanto movement occurring at farmers' markets across the country and many organic farming groups joining together to lobby for the labeling of genetically engineered foods. Along with intrinsic concerns for justice from within religious traditions, religious agrarians also adopt justice-based critiques of the industrial food system advanced by ecological agrarians as they articulate their concerns and ideals for just food and farming.

For many farmers around the world, "The business of farming is, at the end of the day, constrained by the playing-field of the market" (Patel 2007: 7). Such constraints can and do lead to tragedy: loss of farms and livelihoods, loss of regional culture and social capital, and increasingly loss of lives due to suicide out of shame and frustration because the farmers are unable to repay debts.

Those sympathetic with the industrial farming system, while lamenting such "externalities," argue that because farming is a business, those who succeed are better able to handle the dictates of market capitalism. For these supporters, the market eventually will lead to efficiency in food production and development of technologies that save labor and lead to greater productivity.

We see here a war over narrative tropes, with religious agrarians siding with the analysis that "Large corporations are very reluctant to cede their control over the food system" (Patel 2007: 11). Although the division between sustainable and industrial farming is one of metaphor, narrative, and ideas, it equally includes political ramifications codified in legal regulations, agricultural policy, and international commerce. These combine to create material repercussions in the physical bodies of farmers, farm animals, plants, and soils.

Religious agrarians seek ways to farm that honor creation, health, and locality. In this striving they are working to generate food systems that are just for all involved and provide for a living wage for all involved. They view justice as a bottom-up concern, beginning with soil organisms, farm animals, farm laborers, farmers, farm communities, and consumers of local sustainable foods. From this perspective, agribusiness corporations such as Monsanto are largely seen as the "boogey man," creating an uneven, toxic, unjust playing field that harms everything seen to be good about sustainable farming and agrarian lifeways.[1]

Given these beliefs, many religious agrarians are motivated to model, support, and practice sustainable alternatives. In this way, "Social movements [such as CSAs and farmers' markets] provide tangible support and help to communities in need. But they also provide hope, and the promise of change" (Patel 2007: 34). Given the severe ills of modern-day agriculture as seen from the perspective of sustainable agriculture supporters—loss of genetic diversity and topsoil, the poisoning of waters, epidemics of obesity and malnutrition/starvation, and the corruption of democracy by agribusiness interests, to name some of the more common claims—for those in the sustainable agriculture milieu, it is imperative to develop and embody healthy, just alternatives.

Religious agrarians share these same concerns, pronouncements, evaluations, and critiques but ground them within religious worldviews. Altogether, the need for hope and change are also motivators as religious agrarians attempt to put their values about justice into practice to show that "another world is possible." We have seen this in Koinonia's desire to be a demonstration plot for God. At this demonstration plot, Christian religious agrarians practice permaculture and its mantra-like call for "Earth care, people care, fair share." Indeed, the last of this triumvirate is explicitly linked to issues of justice. Despite shortcomings in practice, similar concerns are found at global levels of sustainable food networks. The result is that tropes, narratives, and concerns for justice are locally embodied in lives of religious agrarians, thereby helping contribute to processes of microglobalization (Cetina 2005; Huey 2005). At the same time, they are simultaneously spread at global levels via modern technologies where they help form international lived networks built on the shared premises of agrarian concerns. This is seen in the ethics of permaculture, captured in the globally shared motto.

Although there is a long tradition of religious bodies, institutions, and leaders focusing on justice and ethics (Runzo et al. 2003), it is only more recently that there has been an explicit link to environmental issues. These issues range from overconsumption, to issues of carrying capacity and acceptable human population levels, to genetic modification of food, to loss of species diversity, to the wanton waste of food (over 40 percent in the United States), to social and environmental justice issues (Chapman 2000; Brunk and Coward 2010). Food is an element that unites all of these concerns, so by focusing on issues of justice in regard to agricultural practice and policy, religious agrarians are offering a wide-ranging critique of modern industrial farming, technologies that harm the Earth, and neoliberal capitalism. At the same time, they are reinventing and reimagining their own traditions to make them relevant to various ethical concerns of today.

This is seen, for example, in Judaism with the development of Magen Tzedek. This is a new kosher-certifying agency "founded on the principle that we are what we eat[.] Magen Tzedek is an ethical seal signifying that kosher food has been prepared with care and integrity. Products carrying the Magen Tzedek seal reflect the highest standard on a variety of important issues: employee wages and benefits, health and safety, animal welfare, corporate transparency and environmental impact."[2] The 2009 Hazon Food Conference even held a session about Magen Tzedek, advertised as being a talk about "the aspirations of a burgeoning international movement for sustainable, responsible consumption [that] promotes increased sensitivity to the vast and complex web of global relationships that bring food to our tables." This provides clear evidence that religions are changing in regard to contemporary food issues. The process of change involves adapting and inventing new religious agrarian ethics, ideals, and practices that are able to address the changing technologies, policies, and consumer identity and habits of the industrialized food system; in this case, it is the creation of a new business model and organization, rooted in Jewish ritual practices and commandments.[3]

Judaism also contains deep-seated calls for social justice. When applied to farming issues, this concern for justice helps evidence the process of the renewal and reinterpretation of religion, since religions are social constructions responding to the needs and actions of those who create them. For example, this religious agrarian renewal and reinterpretation is present in the work of Nigel Savage, who on May 31, 2011, sent an email to the Hazon online discussion list with the following passage:

> Nowadays shabbat as a concept stands mostly by itself. But in the biblical narrative it's intertwined with a series of other iterations of seven. . . . The Torah teaches that rest for a person and rest for the land are connected to each other; to do one, we have to do the other. In both cases, we do not "allow" the land or the person to rest;

> we require it. The Torah makes the further presumption—implicit in its day, and being slowly relearned explicitly in ours—that shabbat and shmita necessarily involve concern for the most vulnerable within a society; rest for oneself and rest for the land requires concern for the landless, the day laborer, the widow and the orphan. Today we might add the illegal immigrant, the non-unionized, those without health benefits or retirement plans.[4]

Here is an explicit call for concern about how the lowest in society are treated, where in Savage's hermeneutic Torah and ancient Israelite celebrations and customs speak to concern for the land and the poor. Hazon is updating this concern and applying it to the realities of today's industrial food system and the perceived victims of this system: soils, fieldworkers, slaughterhouse workers, and those working in the agribusiness industry without the benefits of union representation.

This is seen most clearly in Hazon's recent effort to reinstitute the *shmita* year. At the 2013 Food Conference, I participated in a workshop on shmita, hosted by Anna Hanau, who with her husband Naf own Grow and Behold. It was attended by ten conference members, all of us huddled under blankets in a yurt while snow fell outside. After receiving handouts that contained passages from Hazon's workbook about shmita, written by Yigal Deutscher, Hanau (and Deutscher, who was also present) helped walk us through the handout, adding comments and soliciting our own interpretation of various shared passages. Because *shmita* means "to release," and Hazon is about building healthier and more sustainable communities, the facilitators asked us to think about what Jewish tradition has to say about what we see and know about the world. In other words, how do we build the world we want to live in?

They pointed out that for the ancient Israelites, shmita was a year dedicated to shifting energy and perspective based on releasing the land and those in debt of their monetary obligations. Hanau and Deutscher described shmita as a system of giving and receiving fairness within a community, but noted that this is not necessarily a system of equality. They explained that Torah was designed to be a system of laws, ethics, and values about a community of farmers. Hazon, and Hanau and Deutscher via the workshop, is trying to apply the shmita values and ethics to today's world by making relevant ethics of a farming nation from thousands of years ago to Jews today.

The specific passages associated with the shmita year deal with laws of gleaning, planting for the needy, and how communally harvested forms of calories must be shared equitably. One thing they pointed out is that in Deuteronomy 15:11, God explicitly exclaims that the needy will always be present in the Israelite community. Through redaction Hanau and Deutscher explained that this means God did not create a utopia—that Jews have to act on duties for others! Thus, working for fairness is a Jewish duty that has been present from day one in the Israelite world.

The duties commanded by shmita include being willing to make loans but to forgive debts and to share food. On the permaculture walk, Deutscher explained how these and the other shmita laws could be put in place via farming. If Jewish religious agrarians focused on perennial plants and trees and land stewardship, over time sustainable farms can be created that enable the practice of observing shmita commandments. The larger context for this observation, however, is the biblical mandate of rest and release and especially justice and sharing. Hazon is attempting to have contemporary Jews grapple with these shmita insights in today's industrialized, individualistic society—highlighting one example of Jewish religious agrarian concerns for justice.

The links between food choices and justice are also made explicit in Hazon's *Food for Thought*, with a whole chapter devoted to "Food and Ethics: The Implications of Our Food Choices" (Savage and Stevenson 2009). The authors share that "The story of our food does not begin and end on our plate. Food is produced somewhere . . . Jewish tradition has long made the connection between food and social justice, exhorting us whenever possible to share our table with the hungry . . . We were once slaves in Egypt; our memory of our experience of injustice is intended to be a constant reminder to do justice in the world" (ibid.: 93). The memory and consciousness of this experience inspires those in the Jewish food movement to embody the concept of *tzedakah*, or justice. By remembering various Torah and Talmudic passages and teachings about justice, Jews are reminded that "There is a direct connection between agricultural production, social justice and our relationship with God . . . We no longer live in the agrarian world of the biblical land of Israel; yet the underlying issues remain with us today" (ibid.: 94).

These issues of justice and food production are indeed relevant to many of the Southern Jewish religious agrarians I met. For example, the CSA programs in Gainesville and Atlanta donate excess food to regional soup kitchens. In this arrangement, rather than wasting fresh produce, the CSAs organize pickup and drop-off volunteers who bring leftover shares (from people who are out of town or who cannot pick their share up, or who even order an extra share specifically for the food bank) to local soup kitchens.

Furthermore, for Isaac Segal, the concept of justice resonates with his conception of what it is to be a Jew. As he explained, "this is kind of what it means to be a Jew: you lead a good life and you take care of the people and the planet and the animals and the living creatures around you." This understanding of his Jewish identity made Segal excited to see leftover shares go to soup kitchens ("I definitely like the idea . . . I really hate fresh food going to waste") and motivated his practice of joining the CSA.

Like Segal, Noah Gottwald emphasized the strong social justice component of Judaism, stating, "I know that my Jewish upbringing definitely has a social justice component to it. And my environmental consciousness also has a social justice component to it . . . I think that there's an overall consciousness in my Judaism that being good to the environment as well as being good to others . . . are tenets of Judaism. . . . I would say there's no question in my mind that Judaism espouses a social justice outlook on life." For the purposes of my argument, Gottwald explained that this Jewish concern for justice indeed extends to issues of agriculture. In his words, "I would say that the treatment of not just the animals on a farm, but the workers on the farm, is equally important." These beliefs and values have motivated him to adopt religious agrarian practices like joining a CSA.

Meanwhile, Ruth Jacobs explained that she donates excess food she purchases at the local farmers' market to a local soup kitchen in part because "just like Torah and the corners of our field, some of it needs to go to those who don't have" fresh food (*Pe'ah*, from Leviticus 19:9–10, part of the shmita laws). Beatrice Tischler echoes Segal and Jacobs by explaining that the Talmudic prohibition against wasting food, *bal taschit*, inspires her to practice sharing food. She points out, "Throwing away food is a sin."

Tischler's friend Naomi Eisenberg is explicit in her understanding of Judaism and justice. For her, the tzedakah aspect of Judaism has influenced her decision to join the CSA. As she explains, "Tikkun Olam means to repair the world, and that's our job. Our job is to repair the world." This repairing means that Jews have a choice to either care responsibly for the planet and all the life upon it, or destroy it. If they choose to repair it, then Jews are to act with tzedakah, which for Eisenberg, "literally translated I think it means charity, but we take it to mean charity of any kind . . . Giving to, supporting the farmer is tzedakah, and we're commanded to do that." Eisenberg is thus inspired to practice supporting sustainable agriculture because of her desire to support local farmers and she sees this as a commandment given by God.

Such a concern for justice is intricately wrapped up in the work of Daron "Farmer D" Joffe, who left the organic farm he started in Wisconsin and went to California, where he taught farming to youth in a juvenile prison. Joffe followed this experience by beginning a nonprofit to set up gardens in low-income communities, at which time he also began planting gardens in schools and boys homes. As he explains, focusing on justice is "kind of a difficult thing to do from a sustainable economic perspective . . . I run a business. A social and environmentally conscious business. We discount and do as much as we can to support those types of initiatives—food deserts. I've been involved in starting farmers' markets that receive food stamps. I've been a mentor to minority farmers. It's a big part of my passion and my mission."

Farmer D directly links the work to his religious values. When asked what inspires him to undertake work for justice, despite the challenge it places on his ability to earn a living, he responded by saying, "It's my Jewish roots, absolutely. It's engrained in my soul. To practice acts of kindness and to look out for other people . . . My parents instilled in me a strong awareness through my Jewish roots of being persecuted. We have family who were in the Holocaust, Holocaust survivors, so [I'm] very aware and conscious of persecution." Here we see once again that the Jewish value commitment to justice, coupled with Joffe's lived experience of ancestral oppression, inspires him and other Jewish religious agrarians to practice sustainable farming with a focus on justice issues.

Dorothy Goldstein echoes these concerns and this identity and how they influence her approach to organizing the Tuv Ha'Aretz CSA pickup at Congregation Shearith Israel (CSI) in Atlanta. Goldstein explains, "I enjoy Jewish community and certainly enjoy doing service and social justice work as a Jew; that's where my identity is formed." This identity led her to actively build on the relationship she formed with Hazon while living in New York City by beginning a Tuv Ha'Aretz CSA partnership when she relocated to Atlanta. In fact, Goldstein, Rabbi Joshua Hillel, and the leadership of CSI were awarded a grant from Hazon to become one of the first national-level Tuv Ha'Aretz partnerships.

From a sociological perspective, such religious agrarian practices undertaken with institutional support can create community, and they allow people to deepen their relationship with their religious identity and heritage. They can also allow people to talk about the interrelated issues of locality, health, and justice. This has been Goldstein's experience over the years she has run the Tuv Ha'Aretz CSA pickup. The power of numbers available from a religious institution means that for her, "the Jewish community [is able to] put their purchasing power into local family farms." This ability is "very important to me and I think it's important to people who end up committing for several years and who become engaged with the process and find out about it." The more the Tuv Ha'Aretz participants get involved, then the more of "an incredible sense of community and an intergenerational sense of community" forms because arriving at CSI to get a share of organic produce "brings everyone in contact with people across the age spectrum and brings families together with other families and people in conversation with the Rabbis and each other in informal ways that don't happen at synagogue, they don't happen at religious school, they don't happen in our day-to-day lives, but they do happen around the pickup table."

Goldstein continues, this time weaving in the specific justice component of the CSI/Hazon CSA partnership, by explaining that CSI has a partnership with the Atlanta Community Food Bank. In this partnership, each week a Tuv Ha'Aretz member volunteers to bring any leftover produce or shares to an emergency food provider. For Goldstein, this action

> incorporates very strong, universal and Jewish values of providing access—meeting food security needs but also through access to healthy and fresh food, which most emergency food providers don't receive donations of. So those are all opportunities to [work for] social justice through supporting local agriculture and meeting hunger needs; volunteering together; having informal avenues of conversation; participating in some educational programs; and giving the Rabbi a platform to do a lot of both formal and informal teaching about the environment.

These various components are united under the banner of supporting local, healthy, justly produced sustainable food while being rooted within the larger Jewish tradition. They are united in the eyes and teachings of CSI's Rabbi Hillel. With a tone of seriousness he notes that on Wednesday nights during the growing season about fifty Jews come to the synagogue to get organic produce from a local farmer. He points out that not that many Jews come on a Wednesday night for *minyan* and to pray evening prayers. This scenario prompted the rabbi to ask, "So which is the more Jewish venture? Which sustains the religious and spiritual life of the Jews of our congregation more? Traditional daily prayers, or tomatoes? Right. I think you could argue tomatoes." Rabbi Hillel interpreted this scenario to mean that "it shows is that these are not ideas and values that are foreign to people's Jewish lenses. To the spiritual way that they see their Jewishness." In fact, Hillel believes it makes sense "to them that their synagogue would sustain their spiritual life through sustaining ethical farming practices. That that's an ethical imperative which Jews should be supportive of and we should speak about as a Jewish value . . . and this is a much higher level of *kashrut*" than a ritually slaughtered animal.

For Rabbi Hillel, synagogue is a place where Jewish values and teachings are discussed, challenged, and put into practice. He supports Tuv Ha'Aretz because it allows the members of his synagogue who are CSA members to practice their Jewishness. As seen, this Jewishness has a strong ethical component that "sustains . . . spiritual life." For Rabbi Hillel, Wednesday CSA pickups are a time when Jewish ethics, values, and identity get put into community practice, not only in minyan prayer but in embodying a higher level of kashrut.

An ability to put values into practice is an important component of personal identity and happiness. Many of the people I interviewed, observed, and worked with while researching this book expressed joy, satisfaction, and a sense of meaningfulness as results of their participation in religious agrarian practices. For many, this joy was partly because they believed they were able to put their religious values into practice while also doing something they believed is positive for the environment and their local community.

To paraphrase Aldo Leopold, to receive an ecological education about the interconnectedness of life leads to the realization that we live in a world of wounds. For many, these wounds are felt at an emotional level (Roszak et al. 1995). Thus, one of the results of developing a worldview that is influenced by environmental and ecological sciences is a concern for the Earth (whether this is seen as a divine creation or not) and the desire to act, personally and politically, for the protection of the environment. The wounds can especially creep in when a person feels overwhelmed and burns out because most metrics of environmental health are in steady decline and continue to get worse. Therefore, by participating in religious agrarian practices rooted in religious identity and community, religious agrarians are able to act in productive ways for justice, health, and local environmental health.

In their lament about the failures of the environmental movement (broadly speaking) to slow the many metrics of health, Ted Nordhaus and Michael Shellenberger opine, "One of the problems . . . is that while public support for the environmental agenda is broad, it is also frightfully shallow. This has implications not just for the challenge of passing legislation but also for the inseparable challenge of building a movement" (2009: 202). They argue for the need to create an environmental politics that motivates and inspires people to become environmentally active, both in practice and in financial commitment to local and global nongovernmental organizations (NGOs). This argument is built on their analysis of social science literature, which suggests that people are happy and engaged when they are in flow, serve others, interact with others, and live a meaningful life (ibid.: 204). Precisely for these reasons religious agrarians, and the larger ecological reformation occurring within religious traditions globally, are a leading edge in the changing landscape of the environmental movement. In Nordhaus and Shellenberger's words, religious agrarian practices allow for people to flow, serve others, interact with others, and live a meaningful life.

By participating in lived practice (being in the "flow"), by serving others due to concerns about justice, by participating in community, and by being grounded in a cosmology of meaning, religious agrarians are actively challenging the industrial food system and are helping lead the way toward a new way of putting religious and environmental values into practice. This religiously based practice helps heal inner and outer wounds that attend to seeing creation being despoiled by human activities like industrial agriculture. This active mix of challenge and healing applies equally to Jewish and Christian religious agrarians.

We have seen that Judaism contains a strong ethical component. The same strength of ethics holds true for the values and practice of permaculture, as well. During the 2010 Permaculture training at Koinonia, Chuck Marsh shared the core ethical tenets of permaculture. These are care of the Earth, care of humans, care of the community, and respect for the intrinsic value of all living beings. In this values-based worldview, care of the community entails fairly sharing the products of the Earth, where such products should ideally be grown and distributed along the design parameters of permaculture. Therefore, a native value of the permaculture system is to share with those around you.

Such sharing and compassionate commitment to justice has been central to the vision and mission of Koinonia going back to its beginning under Clarence Jordan. Modeled after the radical and egalitarian sharing of the early corporate church as described in the book of Acts, Koinonia has always been a place of voluntary poverty and simplicity that has challenged the iniquities of society. Food and fellowship have been central to this challenge from day one, from working across color boundaries in the fields to sharing food across color boundaries in the community's cafeteria. This spirit and ethos is central to the hospitality and education of Koinonia so that the fair share tenet of permaculture is easily adopted into a culture and institution already used to campaigning for issues of justice. As Sylvia Castle explained, "I'm more concerned probably with peace and social justice [issues over environmental issues] although the neat thing about permaculture is it tries to incorporate all [of them] so it [is] more of a holistic thing than just organic gardening."

Koinonia members have protested the US Army School of Americas, campaigned for peace in a post-9/11 world, collaborated on immigrant rights with

Jubilee, and created a monastic community on the margin of empire. In these activities, justice has been an explicit value. This concern also has led Koinonia to forge networks with business partners who share the same concerns. It is represented in individual lifestyles and actions of community members, seen from community member Nashua's solidarity with Latino immigrants to community worker and preacher Norris's effort to start a food bank for impoverished African Americans in the Americus (Georgia) region.[5]

The concern for justice motivated Ben Peters to become a community member. Prior to joining Koinonia, Peters had done justice work in Philadelphia but realized life in the city was not spiritually or physically healthy. He realized, "I want[ed] to be someplace [where I] could live in a way that didn't require other people to be oppressed or to be suffering. I feel like in the city you use an infrastructure that could be built with unfair labor [and] your whole way of life is based on . . . extracting stuff from the environment. I know in a rural place it is too but it's, one, it can be not as much, and two, it may [not be as easy] to see so I'm happier." We see again that this lifestyle choice is motivated by values Peters developed through lived experience. Whether seeing the chemicals Koinonia uses on its pecan trees after moving there or realizing the eco footprint (and social footprint) of living in cities, Peters's concern for the Earth and for justice obligated him to change his lifestyle. This realization prompted him to move to Koinonia so could learn about and implement permaculture farming techniques while honoring his commitment to live in intentional community, all while striving toward a Christ-centered life.

He explained his own progression in the following words: "I guess my whole life is caring about a certain small group of religious minded people up until [I realized that] people need to be treated fairly no matter what their religion or whatever and its kind of just expanded to all creatures, you know whether they be animals or plants or rocks or whatever." When asked if this motivation to care comes from his faith, Peters responded by saying, "I guess when I read the Bible, like when I read Jesus in the Gospels, it was a different picture than when I had as a kid, so I went from being very sectarian to being, I felt like what Jesus was saying was that he was calling a certain group of people to come follow his teachings to bring about the kingdom of God, which is a term I believe means a world of right relationships." When asked if Koinonia provides a place where he can work on right relationships, Peters responded, "Yes, [because] I wanted to live in a way which didn't require other people to suffer, or at least try to minimize the ways that I cause people to suffer, so I heard of this place [and] I decided to try it out." His motivation was his desire to "live in a way that didn't cause suffering" but this meant that "to do that I had to grow my own food or know people who grew my food so I wouldn't be reliant on the government to subsidize corn which is causing Mexicans to lose their land because they can buy our corn cheaper because of the subsidies we have and the opening up of trade agreements."

Like many religious agrarians, Peters's religious values (his interpretation that Jesus preached an ethics of right relationships) influence his views about the Earth and justice and have compelled him to put into practice his ideals. As he explained, "I think that when Jesus taught, what Jesus was aiming for was equality and you can't . . . I think inequality produces a lot of the world's problems." This interpretation of Jesus's teachings was also held by Clarence Jordan and is one of the reasons all of the permanent members of Koinonia take on vows of voluntary poverty—their room and board is covered by the community, and they earn a little bit extra each year, but ideally they share what little profit is made from sales of pecan products and work to challenge the world's inequalities.

In Peters's analysis, right relationships include our relationships with food: how it is grown, who grows it, and the governmental policies that influence this. By living at Koinonia and putting into practice agrarian and permaculture farming techniques, he is able to live in a way that is consistent with his food justice values.

Stephanie Paterson is equally motivated by justice issues, and before her visit to Koinonia, she was "really into solidarity with the poor and taking care of the homeless." She feels that God is the only ruler over anything, as he is the Creator talked about in Genesis, and humans are "enslaving plants and animals and the Earth" when we should be living in harmony with God's creation.

When asked about why she is so concerned about justice and equality and caring for people and the Earth, Paterson answered, "I believe that God is in every living thing. God is in the soils and God is in the rocks and in the water and in the sky." This understanding of God inspires her to live a life where she is "taking care of people, first and foremost, like that's the most important thing to me is taking care of people . . . And being obedient to God. And I've never really felt as intimate with anything as I do with the Earth. Like that's just a part of who I am."

Paterson extends her desire to abide in caring relationships to the Earth. "I don't think like we're on the land or we own the land. I really believe we are the land, we're just as much a part of it as like plants are and animals are." She recognizes that these are not mainstream Christian views, but nonetheless she is comfortable and happy to call herself a Christian and she feels she is doing the work of Christ. She readily admits that her understanding of Christ was influenced by her exposure to the religious agrarianism of Koinonia. In fact, she claims that

> I didn't think it was that important until I came here and now I realize that [to] love your neighbor as yourself and [to] love God—like if we love the Earth, [then] we're loving God. And if we take care of the Earth, we're going to be able to provide enough for our neighbors . . . if we practice permaculture and organic farming more and more and we begin to live more gently on the earth and more intimately with the earth, we're going to solve all these problems. You know, we could, we can make it so that no one's hungry anymore and so that everyone has a place to live.

Paterson's strong identity with God and the Earth motivate her concern for justice and caring relationships. We see her interrelated religious agrarian concerns for locality, health, and justice, which for Paterson are explicitly linked to her Christian identity and understanding of God. This was reinforced at the end of our conversation, when she volunteered, "I believe that living more sustainably is our solution to poverty and I think Jesus had a big heart for the poor and that was really what he talked about was taking care of the poor and the downtrodden."

Another community member shared how ethical business practices are a key component of Koinonia's business model. These practices include using Fair Trade chocolate from the Divine Chocolate company (their chocolate was available at the 2013 Food Conference) and using Fair Trade coffee beans they buy from a local business, Café Campesino. Koinonia is looking into becoming a certified Fair Trade business as well, meaning they would be recognized for paying their workers living wages and offering health care. Katie Taylor is in charge of maintaining and deepening these business relations, and she explained that using Fair Trade products is "part of our value system" as a community. As she explained,

> We kind of do mutual marketing and I'm actually very interested in those kinds of relationships. I guess I see marketing not so much from a competitive standpoint but from more of a grassroots standpoint. If we're promoting them and they're promoting us, and we're diverting people away from the big corporate, "get rich quick" folks, and more money is going into our pockets, we're all winning.

> Everybody [wins,] right down to the people growing the products and making the products. So that's a big thing for me and my philosophy.

Once again, individual values influence lifestyle choices. In this case, the values are of mutuality and cooperation, and the choices are sharing and a bottom-up economy based on justice and fairness. In this alternative economic model, small business owners and workers who produce sustainable, fair trade products are privileged over the profits of corporations.

These values of sharing and protesting exploitation of workers and soil also motivate Patrick Smith, who lives with his family on the edge of Koinonia's property and walks to work in the fields during the week. One of the practices he is involved with at Koinonia is dumpster diving, which he believes "fits pretty well with the idea of reducing waste." He points out, "there's so much waste in our culture and if we can funnel some resources that are headed to a landfill and feed hogs or feed people or chickens or what have you or turn it into compost even and possibly even recycle some of the containers that this stuff comes in, [then] I think that can be seen as ministry as well. And fits in with permaculture design." In Smith's religious agrarian practices, the waste of industrial culture becomes sacraments that can be culled from dumpsters and turned into inputs for Koinonia's fields and animals. By reclaiming the wastes of "empire," Koinonia can honor the labor of others in that rather than this labor ending up in a landfill, it becomes an object of eco-ministry that helps the health of the community.

Overall, because of its own storied history and its relationship with Habitat for Humanity Koinonia receives thousands of visitors each year. Their catalog and products are mailed to thousands each year. Therefore, the community is able to impact Christian (and non-Christian) food consciousness far outside the boundaries of its few hundred acres, and this impact includes a very strong vision of peace and justice.

In the same vein, as the leading voice of the new Jewish food movement, Hazon is able to shape in profound ways the progressive edge of Jewish food choices. Both groups are part of the "fair food revolution" (Hesterman 2011) and are helping shape religious agrarian discourses about and understandings of just, humane food and food choices. By participating in conferences, giving talks, maintaining websites, and producing publications, and by participating in lived networks, the religious agrarians I encountered are already acting on the realization that "we need to not only eat local or plant gardens but start engaging as fair food citizens to alter the larger policies and institutional practices that are driving the [food] system in its current direction" (Hesterman 2011: 132).

JUSTICE IN THE FUTURE

The history I have been outlining suggests that religious agrarians will only become more visible, organized, and vocal in fair food discussions and policy proposals in the coming years. Given the historical interplay between religion and politics in the United States and given the religious agrarian worldview, it seems prudent to predict that the concern for justice will continue to mobilize and capture attention within religious subcultures in North America. Such attention will most likely lead to continued support for or participation in Fair Trade networks; providing soup kitchens with local, organic produce; and active lobbying at local, state, and national levels for laws that support workers' rights and/or sustainable businesses. Because these activities are value-driven by members of religious institutions, they will most likely become persistent efforts, whether within a religious group or across ecumenical lines.

The search for justice in food production will also grow largely because if the industrial food system fails in its promises of feeding the masses healthy foods and appears instead to be brittle and flawed (a real possibility during this century of ecological limits), then alternative cultural narratives about farming will percolate into mass consciousness. Many of the religious agrarians I observed are evidence that this shift of environmental agrarian concerns into popular culture is already under way, as many have read Berry and Pollan and seen *Food, Inc.*

If perceived limits of health and justice are crossed and experienced, then the United States will most likely undergo an even more profound dialogue about farming and food choices. Religious agrarians will be situated to lead this dialogue, especially within religious communities and networks. Such a dialogue will most likely recognize that

> Farming offers us a comprehensive view of the larger culture that cuts across every concern we have in creating a sustainable society: poverty, the environment, racism, even finding and cultivating a meaningful life rather than a merely prosperous one. Agriculture is one of those great cultural intersections where environmental concerns and social justice issues, including the economy, come together and play themselves out for good or ill. . . . Agriculture, not the usual economic markers of our prosperity, is the keystone for the whole culture. (Holthaus 2009: 118–19)

These sentiments are already on display in religious agrarian values and worldviews, with the concern for justice interweaving itself with concerns for health and locality. Such interaction is seen in the following sentiments expressed by Rabbi Hillel as we discussed food issues in his office:

> The land is local! And it's going to have to stay local. You can't move land around. Globalization turns out to be very significantly limited and restricted in its capacity to do certain things. It can do certain things, can't do other things. And the risk that we expose our farmers to, financially, in order for us to have cheap available food [is unacceptable]. Which is just cutting off our own future. You're stealing from yourself to have the cheap and easy food that jeopardizes the future of the farmer, because if he's not there, I can't grow my own food! I literally, my life depends on the farmer growing the food that I eat. I can't do it. Nobody can really sustain their own life. And self-sustaining farming is very complicated! It turns out you can't do it. You've got to have an interconnectedness of farms and people and communities.

It is precisely this interconnectedness that motivates religious agrarians to put their values into practice. For them, this interconnectedness occurs in three interrelated sets: the first is an interconnectedness of farms, people, and communities. The second is an interconnectedness of locality, health, and justice. The final interrelated set is the interconnectedness of the divine/God/Yahweh/Jesus and creation. Here we see that perception and values drive practice and behavior in regard to striving for a religiously inspired sustainable life. We also see that lived experience, whether smelling the stink of concentrated animal feedlots or reading signs that warn about the dangers of just-sprayed industrial farming chemicals, equally motivate religious agrarians to practice sustainable lifeways. This iterative process of emplaced and embodied value and meaning making is thoroughly grounded in lived networks, religious identity, and a holistic, theocentric worldview, while giving credence to the claim that "We are always already involved in political projects of persuasive truths" (Bauman 2014: 150).

What might engaging in food justice create for religious agrarians, and thus religion broadly, in the coming years? As I have been suggesting and as ecological agrarians have been arguing, the industrial food system is doomed to failure. Settled agriculture itself going back 10,000 years as a whole might be intrinsically flawed, but once humans began applying chemical fertilizers crafted with natural gas and oil to fields, the stability of this method of producing calories became very tenuous. This is a key reason Wes Jackson is working on creating perennial agriculture at the Land Institute—to correct this millennia-old shortcoming made all the more portentious by the use of nonrenewable fossil fuels. When this brittleness is grafted to concerns of justice built on the vector of religion, and if religious practitioners are engaged in political projects of persuasive truths, then food justice issues become a place to "think like a planet" and build a planetary ethic (Callicott 2013). Yet significantly, this process of thinking like a planet also entails, at least for religious agrarians, fidelity to the local, where the local is typically seen as sacred or religiously significant. Last, thinking like a planet entails laboring for the generation of health.

If religious agrarians think like a planet, then their justice component should not be underestimated in its ability to mobilize and inspire. As Larry Rasmussen explains, "Prophetic-liberative ethics" (2013: 329) are steeped in a critique of power, and furthermore demand action in subverting power. Religious practitioners motivated by these ethics are actively engaged in naming, unmasking, and engaging the powers that structure society so that these powers (i.e., institutions) can be redeemed (Wink 1999). This is what Rabbi Noah Farkas meant when he stated at the 2013 Hazon Food Conference that the Jewish food movement needed to mature into adulthood, where this maturing would be signaled by a more strident and vocal critique of the industrial food system and the power structure that enables it. Thus, religious agrarians offer a corrective to Rasmussen's claim that "many prophetic-liberative strands [in religions] have yet to overcome the anthropocentrism of their social justice traditions" (Rasmussen 2013: 361).

Precisely because religious agrarianism is influenced by ecological agrarianism and takes the local, soil, animals, workers, and a divine creation seriously and as important in ethical decision making as the self, there is the potential of creating "creation justice" (Rasmussen 2013: 361). In this effort, influenced by mixing concerns for locality, health, and justice, religious agrarianism is helping embed the reemergence of place in the United States. I return to this theme in my conclusion, teasing out implications for the study of religion, and the procurement of human calories, more broadly, in my final chapter.

CHAPTER SEVEN

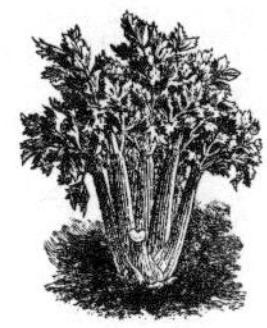

CONCLUSION

A HARVEST OF IDEAS

A CHANGING LANDSCAPE (OF FARMLANDS AND RELIGIOUS STUDIES)

Over the previous three chapters I explored emergent themes from my research that form the basis of what I call religious agrarianism. The repeated references to, concerns for, and attempts to articulate religiously grounded ideas of locality, health, and justice define religious agrarianism. Furthermore, various lived networks within which religious agrarians participate, help shape, and in return are shaped by influence this definition. Consistent with a lived religion methodological approach to studying religion, I explored the themes of locality, health, and justice. Theology, scripture, and orthodox teachings of Judaism and Christianity are relevant to the religious agrarians I met insofar as they influence their lived practices and understandings. Yet religious agrarianism is significantly a bottom-up phenomenon, shaped by the values and practices of religiously motivated individuals and groups who point to and cite scripture and orthodox teachings as needed but who are not defined or bound by them. Rather, religious agrarians choose what is most helpful as they articulate an ecological and agricultural sensibility that is present in their lives and often informed by their experiences.

These experiences help shape individual and group approaches to the environment, agriculture, and thus religion. Such an iterative process leads to selective interpretation or reinterpretation of teachings, doctrines, and scripture and the formation of new practices (like a CSA pickup at a synagogue) so that these become elements in an emergent religious agrarian worldview. In this way, religious agrarianism is both a social construct—reacting to the politics, economics, environmental footprint, and cultural workings of industrial and sustainable agriculture models—and a lived religion that morphs into lived networks that reinforce its social construction.

In this concluding chapter, I offer comparison and contrast in how Jewish and Christian religious agrarians in the United States conceive of locality, health, and justice. I return to network theory, environmental philosophy, and religious studies theory to finalize my analysis and argument about religious agrarianism and its emergent role in the religious landscape of the United States.

REVISITING THE "RELIGION" IN RELIGIOUS AGRARIANISM

Let us briefly revisit, compare, and contrast how religious agrarians use, shape, and reinvent aspects of their faith to validate, justify, and inspire their lifestyles. In this way, we see how the emergence of a religious agrarian worldview is a unique and recent development in North American religious history while it also signals the continued development of "green" religious sensibilities.

Many analysts assert that religion is central to the worldviews that underlie sustainable agrarianism and the recent emergence of "foodie-ism." For example, Eric Burkett claims that "Religion, of course, takes many forms, but its most interesting form to date is food. Many folks, it seems, have embraced food, or food activism, as a new religion."[1] I do not agree with Burkett's argument that food activism is a new religion largely because the alternative food movement is too broad, lacks a centralized institutional structure, does not enter into talk about the supernatural, and lacks defined rituals. I sympathize with his recognition that the modern foodie landscape (and its sustainability aspect) is peopled with many who have a zeal and passion for all things food. This almost "born-again" passion leads such people to crusade for food issues and enact foodie lifestyles. If anything, however, the religious agrarians I met place their food concerns within a larger container: their own religious traditions.

My research also challenges an analysis offered by the leading agrarian writer Gene Logsdon, who asserts that "Many modern agrarians are wary of institutional religion, like the ancient *pagani* were, and for the same reason. They view it as a product of organized urban power trying to gain control over them—trying to colonize them for its own benefit" (2007: 303–4). Logsdon's research and experience shield his gaze from seeing that there is an emergent, modern form of agrarianism: religious agrarianism. Clearly the modern agrarians introduced in this book do not hold antipathy toward religious systems. In fact, most members of Hazon *are* residents of urban centers of power, and they use this position to help local sustainable farmers.

In terms of locality, the urban Jewish research subjects were interested in fresh food and taste. Many come from a culinary background where their parents cooked traditional Jewish meals. Quite a few grew up with backyard gardens and missed the taste of fresh food they enjoyed in their childhoods. For the majority of the urban Jewish religious agrarians I encountered in the South, concern about the local meant the ability to enjoy fresh, seasonal food while cooking meals for friends and family and Shabbat dinner. This is not to imply that the only concerns about the local were for fresh, seasonal food. Rather, this partly signals that members of Tuv Ha'Aretz tend to be urban residents who join the CSA to get fresh food in a uniquely Jewish community setting.

When other values and motivations about concerns for the local appeared in my research about Jewish religious agrarians, it was contextualized within a few key concepts that often overlap with values about health and justice. One was *tikkun olam*, or the Jewish mandate to heal the world. Another was *bal taschit*, the ethical mandate to not destroy or waste. Another motivating factor was interpretations of Jewish scripture and teachings that highlight the agricultural past of the Jewish people. Some Jewish religious agrarians draw inspiration from the many blessings and seasonal holidays surrounding food that they locate within their tradition, especially in the Torah or Talmud. By participating in sustainable agriculture—whether directly or through a CSA membership—these Jews invest renewed meaning into this ancient part of their religion.

Some Jews are motivated by the human responsibility, given in Genesis, to be stewards of the Earth. They interpret this mandate as a duty to care for local land bases by farming in healthy ways, as seen in the work of Daron "Farmer D" Joffe. Many are also motivated by a desire to support (and even start) local businesses, although this is a more secular concern (unless those started are sustainable farms, eco-camps, or kosher slaughterhouses that serve Jewish clientele, as seen in the vibrant lived networks that constitute Hazon and its partnerships).

In comparison, members of Koinonia are much more concerned about the environmental health and ecological well-being of their immediate community. This is largely because these Southern Christians live and work on the land that provides their food and that they are to steward. This land is outside their windows, and they walk across it every day on the way to morning prayer at the chapel, the way to work at their various duties within the community, and when taking a leisurely stroll along the peace trail through the pine forest. Many of them have also participated in permaculture trainings so they bring a permaculture view of intentional design to their vision for what their locality will resemble; this vision is codified in the operating protocols and guiding parameters of the community.

However, specific Christian beliefs, theologies, and concerns regarding "the local" motivate those involved in Christian agrarianism at Koinonia. These include creation care and holding an incarnational worldview, meaning that by seeing the Earth as being both the creation of and home to God, there is a Christian obligation to care for the Earth. For many in the creation care movement, this charge to care for the divine creation is seen as a blessing and a potential vehicle for grace rather than a task to soberly undertake in fear.

Members also view Koinonia as a demonstration plot where God's ideal of love and compassion are put on display. Included in this demonstration plot is Koinonia's attempt to forge and articulate new human–divine relations by permaculture and other agrarian practices. Members also hold a Christ-centered commitment to radical poverty and sharing that is inspired by the life and teachings of Jesus (as well as Clarence Jordan and contemporary New Monastics), who is viewed by the community as being either the savior or the human being par excellence. Last, members of Koinonia believe that we are members of and participate in God's kingdom, so how we treat that kingdom matters. This insight is shared by Regan Sutterfield, who writes that the

> agrarian view . . . is oriented to the redemption of this place and the bodies dependent on this place. Here we move closer to historic Christianity's expectation of time, where we begin with a fall and have been working through the history of the world toward redemption. Our work is not oriented toward some heaven in the sky, but a redemption of the world that God so loved . . . the good farmer is not called to make her farm as productive as possible so that she might feed the world. She is called to be responsible toward its gifts. (2013: 160)

Koinonia is thus one of many vibrant voices articulating an emergent religious agrarianism that cares for place because that place is a creation of God.

Even here, Koinonia adds a unique perspective. For example, while morning services and noon mealtime prayers did refer to nature (for example, Joel Taylor, Koinonia's lead farmer, once read a passage from Wendell Berry during a morning prayer at the chapel), most prayers and passages I witnessed referenced concerns about peace and reconciliation and about staying true to the vision of Koinonia. Few directly cited passages from scripture, but many believed that they were carrying the vision of Clarence Jordan forward into the twenty-first century.

Many at Koinonia cited Jesus as their moral and spiritual exemplar, especially in regard to justice. However, after our initial interview and after working many hours together and talking in the permaculture fields, Paul Robinson approached me with a scrap of paper that contained two passages from the New Testament that influence his view of locality and health and that together help motivate his actions. The first was Colossians 1:15–17, which states that all things, including all things on Earth, were created in Jesus; the second was Hebrews 1:3, which shares that the Son radiates God's glory. For Robinson, these passages inspire a life that is dedicated to serving, sustaining, and keeping healthy local ecosystems, as these are created by God and serve as a reminder of the radiance of Jesus and God. Others were inspired by the Genesis vision of stewardship, while the Sermon on the Mount motivated others, although this influenced visions of justice more than locality.

In regard to religious agrarian concerns about health, members of Hazon were influenced by the concept of *tikkun olam*, which directly speaks to healing and health. Many of the Jews I spoke with see this healing to be an interconnected healing built on healthy food, healthy local economies, healthy bodies, a healthy Jewish identity and community, and healthy, sustainable human–Earth relations built on ecological agrarian farming practices. Others actively enacted Jewish rituals that are designed to bring a Jew into presence with their food while also giving thanks to God. They felt that these practices, and the emotions precipitated by them, were heightened because the food being consumed was sustainably grown. Others believed in the concept of hospitality and enjoying "slow food" (concepts shared with Koinonia), which related to the health of bodies and the salubrious effects of sharing sustainably grown food at communal meals. Finally, some Jews explained that they felt the need to reexamine and reimagine what it means to be kosher in an age of concentrated animal feeding operations (CAFO) because how kosher food products are raised and processed directly relate to Jewish concepts of health and justice.

In comparison, the Christians I studied are concerned about the various aspects of health because they believe we are to care for God's creation and are not to harm it. They believe they are called to model holistic farming to visitors. In this respect, those at Koinonia believe they are following the community's unique institutional culture of demonstrating new, harmonious ways of being in society, with this harmony now extending to human–Earth relations via permaculture. Last, both groups believe that they are commanded in some way—through revelation, scripture, and/or teaching—to live in harmony with God's creation.

Meanwhile, Christian and Jewish religious agrarian concerns about health are influenced by ecological understandings of soil and physical health. In other words, members of both religions agree and believe that the chemicals used in industrial farming are poisonous to the Earth and to human bodies, evidencing the impact environmental science is having on religious production in subsets of North American religions. In this secular reading of various prosustainable agriculture literature, studies, and media, religious agrarians in the United States are entirely aligned. Moreover, in their analysis of what this means to their religious identity, they are more synonymous than divergent: many view these poisons as being an affront to a divine creation or an affront directly to God. The groups share similar secular critiques of our at-large societal health: North Americans consume too much, consume too much unhealthy food, and do not make an extended effort to support local, small-scale family farms.

Finally, the religious agrarian groups who form the basis of this book were extremely motivated by concerns about justice. For those at Koinonia, issues of justice have motivated the community since its founding. This concern is still present and extends to how all those involved with industrial agriculture are treated (workers, animals, soils) and even motivates their purchasing decisions. It inspires them to be members of the New Monastic movement. Many point to the teachings and examples of Jesus and Clarence Jordan as their motivation for holding concerns about justice and how justice is a part of diet and food choice. This concern is strongly echoed by the Jewish religious agrarians I met, who almost all claimed that Judaism takes justice as a central concern. For them this concern is found in scripture, sermons, and teachings, and for many the message was imparted by family members during formative childhood years.

While at the 2009 Hazon Food Conference, Rabbi Jacob Fine, assistant director of Hillel at the University of Seattle, presented an articulate synthesis of the various Jewish religious agrarian concerns for locality, health, and justice that neatly summarizes much of my research. This occurred during a Friday morning talk about Jewish food ethics where the goal was to build on Jewish tradition to frame a contemporary Jewish food ethic. He began by pointing out, "We are living in a new age if we're talking about *kashrut* including labor, environmental issues, and animal rights issues." This is a consistent theme in my research: many religious agrarian Jews recognize that there are new food issues that the tradition must address, and not least is the need to update what it means to keep kosher in today's industrialized food system. The concern for keeping kosher motivates Jewish agrarians and is conspicuously absent in Christian agrarian circles, so this is one of the biggest differences between the groups.

The whole process of living a Jewish life guided by Jewish ethics, *mitzvoth*, commandments, and laws is unique to the Jewish agrarian worldview. Jewish agrarians and food activists tended to warp environmental agrarianism into the woof of the much longer, larger, and more meaningful Jewish identity in which they were raised. For Jewish agrarians, this identity stretches back to the earliest patriarchs and the agricultural cycles of the earliest Israelites and proceeds through the rabbinical tradition to today, where kashrut and Torah (and for Hazon, *shmita*) must deal with health, justice, and environmental issues brought about by industrial agriculture.

Fine identified ten ingredients that a new Jewish food ethic should include or on which it should be built. The first ingredient is *tza'ar ba'alei chayim*, avoiding cruelty to animals, which relates to all three religious agrarian tropes. The next two ethics are interrelated; they are *lishmor et ha'adamah*, or the injunction to protect the Earth, and *bal tashchit* (do not harm/destroy), a concept explored in prior chapters and readily used by other Jewish religious agrarians.

The fourth ingredient of a modern Jewish food ethic is *lo ta'ashok sachir ani v'evyan*, or the injunction to not oppress laborers, evidencing a concern for health and justice. This ethic is related to the fifth and sixth ingredients, *kavod habriyot*, "providing dignity for all people," and *ayn mahazikin yidei outei aveirah*, which means "one may not aid and abet a transgressor" by buying products from "Big Ag" corporations.

Number seven is *hillul Hashem*, or desecration of God's name. This ethic is related to abstaining from genetically engineered ingredients and not taking shortcuts with God's laws (i.e., not corrupting what it means to be kosher). Number eight is *lifnim mishurant ha-din*, "going beyond the letter of the law," creating a food system that is beyond kashrut, that is beyond organic, that is beyond just, and that is beyond healthy. Finally there is feeding the hungry based on the Torah (gleanings of the field for the poor and donating 10 percent of a harvest to the poor) and concepts of *tzeddek* (justice, righteousness); the final ethic is *shmirat haguf*, or "protecting one's health."

We see here many of the motivating values, concerns, and teachings extant in Judaism that are used to influence and shape a nascent US Jewish religious agrarianism. Many of these values emerged over the course of researching this book and were shared in chapter 3's literature review. These Jewish concepts of justice, the teachings and laws from the Torah and the rabbinic tradition, and the agricultural bedrock of the early Jewish identity contribute to the emerging

Jewish agrarianism in the United States of which Hazon is the leading voice. As Fine put it, "This is an exciting moment to be Jewish, and us being here [at the Food Conference] together is one of the reasons. When looking back at Jews and food, this group will stand out."

Evidence of the interlocking concerns for locality, health, and justice are also becoming increasingly abundant in various Christian settings and not just at permaculture trainings held at Koinonia. For example, the Lake Junaluska conference and retreat center in North Carolina held a Caring for Creation conference in spring 2011 (Lake Junaluska is part of the United Methodist Church). This was an ecumenical conference with an interfaith component for Christians concerned about the state of creation. One presentation topic was permaculture, and Zev Friedman—who also was a teacher at the 2010 training at Koinonia—taught this module. This reflects the network links connecting agrarian activists and farmers and overlap of these links with religious groups in the South (and in other parts of the country). Other presentations addressed sustainable agriculture, going beyond the basics in greening one's church, faith principles of creation care, creation care and poverty, and triaging the train wreck of climate change.[2] These various concerns are becoming more omnipresent across and within Christian denominational lines, supporting my overall argument that the subset of green religious agrarian concerns are becoming attractive to particular segments of North American religion. Furthermore, the attractiveness of the values relating to locality, health, and justice that permeate the religious agrarian worldview has specific consequences in the form of concrete, sustainable agricultural practices.

PLACE/S, BOUNDARIES, RESILIENCE, AND ECOPHENOMENOLOGY: NEW FUTURES FOR RELIGION

As the practice of religion in the United States continues to unfold in this century of growing climate destabilization (Hertsgaard 2011; LeVasseur 2015b), issues of environmental and ecological health and trauma will become a more pronounced focus. This issue will come to dominate sermons, hermeneutics of a tradition, exegesis of sacred texts, and the creation of lived practices and will present a challenge to orthodoxy and orthopraxy. As humans descend the other side of peak oil great disruption (Gilding 2011) and are forced to live within nature's limits (Hardin 1993), values and practices will change. The social constructionist and lived religion approaches to religion I use in this book expect such changes.

In fact, what I present here reveals that this process is already occurring. This is true especially in regard to sustainable agrarian and sustainable food issues, where this process is greatly enhanced by memberships in lived networks. Values about locality, health, and justice motivate religious agrarians to participate in sustainable agriculture lifeways. These values are directly shaped and influenced by religion and by society, politics, participation in lived networks, and environmental science. It is clear that certain segments of religions in the United States are indeed changing and reconstructing their values and beliefs to live out these emerging environmentally centered values, at least in regard to sustainable food production and environmental agrarian mores.

The fusion of religious environmentalist and ecological agrarian concerns within religious agrarian communities and networks in the United States suggests that these communities will continue to develop religious values and sensibilities that will take seriously the biocapacity of local bioregions. Because the practice of sustainable agriculture engages accepting the ecological limits of God's creation while providing a method of farming that can renew soil and society, the biocultural construction of religious agrarianism will continue (Baker and Winkelman 2010). This process and these constructions will signal a continued shift on the North American religious landscape. Because of religion's role in the United States, this shift will continue to effect US politics, business practices, interfaith interaction, and cleavages and debates within religions. Religious agrarians are at the forefront of this effort to put ecological agrarianism, sustainable agriculture, and religious teachings into practice.

The vanguard theologies, affective emotions (Schaefer 2015) toward the Earth via green cosmologies, and institutional practices of religious agrarians will influence religious ethics and thus the material environment in various ways. One important influence will be the way religious practitioners transform their relations to their local environments. Therefore, I expect that the key religious agrarian theme of locality will continue to grow in importance in the religious future. This is because of the importance local food production will have in a post–peak oil, climate-destabilized future, where growing seasons will shift and the cost of petroleum products for both shipping and fertilizers will become prohibitive. Although our current lifeways are dependent on abundant oil, natural gas, and stabilized weather patterns, the reality is that "both culture and ecology vary from place to place, both are complex, interdependent, and dynamic. Consequently, long-term sustainable production solutions must adapt to local conditions—both cultural and ecological—to be successful" (Kirschenmann and Bird 2006: 316), all while these local conditions are unraveling compared with their historical baselines measured in time scales of human presence. This is even more of a truism when compared to the baseline of the Holocene, the geological epoch we are transitioning out of that was the context for all the domesticated crops and animals our agriculture is based upon, sustainable and industrial. Religious agrarians are positioning themselves to be at the forefront of the role religion might play in local adaptation and long-term sustainable food production solutions in this time of water, oil, and soil shortages and changing weather patterns.

Such an increase in concern for and attention to locality, and how this will shape religious production, signals the reemergence of place. This reemergence also partly calls into question the possibility (desirability?) of a universal and universalized religion and religious narrative. Locality means healthy and just regional foods, regional food cultures, regional recipes, regional landscapes, and regional networks (Nabhan 2008; LeVasseur et al. 2016). It also means a more locally oriented religion. A localized religion will be a lived religion built on lived networks, for lived networks are "phenomenological realities" (Vasquez 2011: 299) where "Meaning, orientation, and intentionality . . . are constitutive of networks themselves. Within and through networks, actors carve out spaces to dwell, itineraries, and everyday routines, drawing from religious symbols and tropes to reflect on and orient their own praxis and to 'sacralize' nature and built environments" (ibid.: 299).

The spaces and everyday routines being carved out by US religious agrarians via lived networks and the translation of values into practice are to be found in farmers' markets, CSA arrangements, and sustainable agriculture fields in communities ranging from Koinonia to Adamah farm. Furthermore, the praxis of religious agrarianism rests on concepts of localism, health, and justice and results in the sacralization of sustainable farmlands and built environments—from compost piles to folding tables holding CSA boxes waiting to be picked up to a pasture of free-range cows.

Religious localism also implies a religious environmentalism fused with ecophenomenology set into a landscape that will shape religious production and praxis. Such a "dwelling perspective" (Ingold 2000: 5, 153) will invariably lead to local skills and local adaptation to local environments, something that religious agrarianism is already beginning to put into practice. This is seen, for example, in the following chain of ideas shared by Rabbi Joshua Hillel during our conversation. When asked what it was about the CSA that appealed to some families and made them think, "Hey, that speaks to me as a Jew, I want to do that," the rabbi responded by saying "I think it's doable. It's an ethical choice which can be made, which has profound impact beyond our own personal life. I think people are looking for choices they can make that have an impact

[and] food is right up there with them." For Rabbi Hillel, "people have taken in and really assimilated the local food message that the long-term health of the earth depends on having lots of farms all over the place. . . . There's a newfound appreciation for the people who feed us and that we cannot sustain a model that every year jeopardizes their very existence. . . . When I [as a consumer] should share the risk."

In fact, joining the CSA means that for Rabbi Hillel there is no risk and the worst-case scenario is that there is a bad harvest of tomatoes. However, he points out that everyone needs farmers to succeed. As he explained, "It's not okay for the local farmer to go out of business, or my food is gone."

The rabbi also added that how a local farmer treats their soil is a motivator for Jews to join the CSA. "A smaller farmer who is specifically dependent on a piece of land may be more likely to take better care of it. But again most economics is short-term planning. People generally don't plan long-term sustainable economies. So it's an unusual person who will have the long-term vision and mind when the short-term opportunity for profit is there, [which is] why most of the economic model is a short-term prosperity." This is because the short-term model does not "have a lot of vision. We just don't have a lot of vision [whereas] Hazon really stands for vision: it's the vision that the world and its food and its farmers and its eaters could be done and run differently."

This vision results in a system that is "better for our personal ecology, better for the sustenance of the farmers and their families and the business which we depend on, and certainly much better for environmental reasons in terms of 'I get the better food, I get good wonderful stuff, the farmer loses the incredible risk that he has to maintain, and spreads that out and spreads the risk and spreads the reward.'" Last, for Rabbi Hillel, "the environmentalism of it is both the extra icing on the cake and it's the fundamental root of what makes this a good idea."

Here the rabbi argues for spreading risk and reward across local religious, culinary, economic, and agricultural landscapes. Religious institutions and individuals with vision can contribute to healthy and just local economies of scale and ecologies of health. For Hillel, this is a win-win scenario for all involved, from local farmers to Jews who can put their environmental ethics and concerns about food into action.

In comparison, Rabbi Elijah Greenberg of B'nai Israel in Gainesville, Florida, is a bit more cautious in his analysis of the contemporary state of Jewish environmentalism. Although he supports Hazon and the CSA alliance with Sandhill Farm, he does not believe Jews have reached a critical mass with their commitment to environmentalism. As he explained, "when 60 to 70 percent of Jews [in a given town] have no connection to any kind of organized Jewish life, can I say, 'Yes it's a growing movement?' [i.e., Jewish environmentalism and especially sustainable food issues.] Yes I'm proud of what we're doing and the commitment we're making [but] I haven't seen the 60 percent that aren't involved flocking in here because we're now teaching this." For this rabbi, the reality is that "as long as the dollar is still king and industry still runs the world," most people are not going to become environmentally concerned.

Nonetheless, Rabbi Greenberg is thankful for his active synagogue and thinks the Hazon relationship and CSA partnership with Sandhill will be beneficial. This is because "for those people who are going to partake in this, for those people who feel connected to the tradition, this enriches and enhances their lives. And to me that's beautiful, that's why I'm here. For me that's the role faith and religion play: to enrich people's lives." In this analysis, Rabbi Greenberg may underestimate the influence of lived networks on identity formation. While a 60 percent increase is unrealistic as long as oil runs industrial lifeways, there is the real possibility of larger ripple effects within the B'nai Israel community over the coming years as more people recognize and are influenced by the enrichment they witness and are exposed to in their peer networks. This is especially possible when this enrichment is grafted on to a rich history of food issues in Judaism that speaks to Jewish identities. Add in the continued exploitation of nonrenewable resources by industrial agriculture and the industrial economy more broadly, and it creates the context for changing food habits and choices.

While both rabbis have realistic perceptions of environmental problems and solutions, they find hope and excitement in Jewish religious agrarianism. If such participation is indeed enriching for those involved, there is little reason to believe this source of enrichment is going to go away anytime soon. My research suggests that for urban Jews, friends network with and tell others about their experience joining a CSA so that more people join. It also suggests that membership in a Hazon-sponsored CSA where Tuv Ha'Aretz pickups are held in synagogues helps generate a rich sense of community that is centered on food and faith. Such an experience can help offset navigating urban shopping centers where fresh organic food is hard to find or is shipped from California. It also allows for a deepening of religious commitment and practice in regard to sustainable agriculture and religious environmentalist values.

Christian agrarians share similar concerns about the importance of locality, the need to support farmers, and the need to generate a just, sustainable economy that privileges the local. For some, this importance is grounded in experiencing the local as being part of a divine creation. Sylvia Castle shared such a sentiment when she explained, "There's something about the farm that is kind of sacred and therapeutic. I don't know if you'd have to necessarily farm it, but I think there's something about growing things, there's something about the seasons, that you just, you do, you feel closer to the Creator. And you feel part of that creation." Therefore, for Castle walking the permaculture fields of Koinonia allows her to feel part of God's creation. She claims that this connection with creation "is lost on many, many people in urban settings and the virtual world. And for many in those situations they long for this peace and to become a part of that." Given this desire, she believes that one of Koinonia's missions is to "sow seeds of transformation" by being a demonstration plot for a Christ-centered religious agrarianism.

Such seed sowing and transformation must be local. It is a physical, spiritual, and therapeutic transformation. Directly participating in Koinonia's permaculture design (or at least witnessing it by walking the fields) places a person within a localized environment that is a manifestation of the Creator. As Executive Director Lisa Jones puts it, "We've got to be true to what we've been called to be. And [to] be as fair as we can and as loving as we can and . . . what's gonna take root is going to take root." Koinonia is following a calling to be faithful, sustainable stewards in a loving and just manner. To do this, member's values and practices must take root at a local level, directly on their campus and in their immediate ecological surroundings. This call to act is both religious and material. For religious agrarians, these two domains cannot be separated, for "in the extinction of agrarian practices we witness not only the loss of a way of life but the erosion of a cultural sensibility that underst[ands] intimately and concretely the human bond with the earth" (Wirzba 2003a: 2).

The intertwining of religion and biological materialism, and the influence of this on lived networks, lifestyle choices, and the social production of religious agrarianism's dwelling perspective, points to the fact that "there is no knowing or sensing a place except by being in that place, and to be in a place is to be in a position to perceive it. . . . Local knowledge is at one with lived experience if it is indeed true that this knowledge is of the localities in which the knowing subject lives. To live is to live locally, and to know is first of all to know the places one is in" (Casey 1996: 18).

The discipline of religious studies needs to take this insight seriously for it underlines the importance of locality. Most directly, understanding the laws of thermodynamics and how trophic pyramids function in generating calories points to the recognition that as evolved animals, all humans are embodied beings emplaced within a living landscape and a culture. For religious agrarians, there is a more ethical way to obtain and consume calories, and the ethics of this path accept local limits while also supporting the effort to better know a local place. Such a shift in religious sensibilities at increasing levels is a major hallmark of the religious environmental present and continued future.

This shift is also built on the growing recognition that our bodies (and our appetites) serve as an entryway into interacting with our local world; that as "*lived bodies belong to places . . . places belong to lived bodies* and depend on them. The lived body is the material condition of possibility for the place-world while being itself a member of that same world. It is basic to place and part of place. . . . Bodies and places are connatural terms. They interanimate each other" (Casey 1996: 25; emphasis in original). The attention to place culminates in culture, for "To be cultural, is to inhabit a place sufficiently intensely to cultivate it—to be responsible for it, to respond to it, to attend to it caringly. Where else but in particular places can culture take root?" (ibid.: 34).

This brings us back full circle to the same views and sentiments of ecological agrarians and the life work of Wendell Berry. Religious agrarians help bridge the gap between a place-based religious ethic and a concern for sustainable agriculture while asking and in part answering Berry's three questions, shared earlier in this book. This phenomenon is historically skipped by literature on religious studies, religion in North America, and food studies literature that attempt to explain support for sustainable agriculture in ways that discount religious values and identities.

Locally based religions can help counter the narratives that guide industrial agriculture, such as this claim by Kenneth Adams: "The development of the several families of antibiotics, insecticides, fungicides, and weedicides, and other farm chemicals have simplified the farmers' never-ending battle against plant and animal diseases, insects, and the other dangers against which he must wage continual warfare" (1976: 88). This narrative has material implications on local soils and farm communities, not to mention the health of physical bodies.

Against this narrative is a religious agrarian cosmology that believes "the most insistent and formidable concern of agriculture, wherever it is taken seriously, is the distinct individuality of every farm, every field on every farm, every farm family, and every creature on every farm. Farming becomes a high art when farmers know and respect in their work the distinct individuality of their place and the neighborhood of creatures that lives there" (Berry 2005: 45). In this cosmology, God knows the feather on every sparrow and the microbe in every ounce of soil. This cosmology also points to a religion of particularity that is grounded in a place and a culture of place. In the permaculture fields of Koinonia, in the CSA partnership between Congregation Shearith Israel

(CSI) and Riverview Farm facilitated by Hazon, and in similar settings the country over, such an emplaced cosmology of the soil is beginning to take root.

However, this return to the local—by both groups, and the focus of our theorizing and research as professionals—needs to equally be placed within the reality of "hyperobjects" (Morton 2013) that supersede and shape us as we live through the end of the world. For Morton, this event began in April 1784 when James Watt generated the patented steam engine and began our addiction to stored forms of carbon (ibid.: 7). As Morton explains, "Our continued survival, and therefore the survival of the planet we're now dominating beyond all doubt, depends on our thinking past Nature" (Morton 2010: 5). We need to move beyond Romantic-tinged views of a reified, pristine "Nature," and adopt what Morton calls "ecological thought" (ibid.: 7). This is "the thinking of interconnectedness . . . It's a practice and a process of becoming fully aware of how human beings are connected with other beings—animal, vegetable, or mineral. Ultimately, this includes thinking about democracy. What would a truly democratic encounter between truly equal beings look like, what would it be—can we even imagine it?" (ibid.: 7). Such encounters are actively imagined, articulated, and thought via a variety of sustainable agriculture practices being undertaken by religious agrarians in the United States.

This praxis of ecological thought and bodily, emplaced making of culture couched in agri-culture generates a gestalt of transcorporality. This means that "by emphasizing the movement across bodies, trans-corporeality reveals the interchanges and interconnections between various bodily natures. But by underscoring that *trans* indicates movement across different sites, trans-corporeality also opens up a mobile space that acknowledges the often unpredictable and unwanted actions of human bodies, nonhuman creatures, ecological systems, chemical agents, and other actors" (Alaimo 2010: 2). A religious agrarian transcorporality infuses hyperobject, transcorporal, hybrid spaces and bodies and fields with concepts of the divine as well, where moving beyond Nature means engaging with and working to embody sustainable lifeways in a divine, articulate, fluid, vibrant creation. In a contemporary interpretation, this can possibly be intuited in the biblical "I am that I am"—the divine as verb, as unfolding action and process, at scales that transcend and include the human.

For religious agrarians, and others involved in the global ecological reformation, this sense of divinity adds another layer to being human. The importance of this layer must not be neglected in our scholarship of human–nature interactions and it must be factored in to our species' response to the Anthropocene as it directly motivates some to try to change their unsustainable practices. That this sense of divinity is grafted on to local places and can be engaged by religious agrarian praxis adds a cautionary corrective to the otherwise astute theorizing of Ursula Heise, who states, "The challenge that deterritorialization poses for the environmental imagination . . . is to envision how ecologically based advocacy on behalf of the nonhuman world as well as on behalf of greater socioenvironmental justice might be formulated in terms that are premised no longer primarily on ties to local places but on ties to territories and systems that are understood to encompass the planet as a whole" (Heise 2008: 10). The engagement with place, connected to concerns with justice and health, helps religious agrarians navigate the religious environmental imagination. If the global, industrial agriculture system, and its contribution to the hyperobject of climate change, is a process of deterritorialization, then sinking fingers into the humus of an organically farmed field is a potent, embodied response.

CLOSING ARGUMENTS

When teaching classes, I often encourage my students to think about course content through the lens of asking "What is at stake?" Responses to this question often hint at or even explicitly articulate issues of power. What is religion? Well, what is at stake in defining this term? Who benefits and loses? What are the political ramifications of claims to authenticity, especially in areas of cultural and religious strife? What are the roles of women or LGBTQ+ individuals in religion? For environmental studies classes: who controls resources? What happens to us when our language reifies concepts such as "resources," and an economy of extraction is built on these fallacies of misplaced concreteness? What are our views of animals and soil and water? What is at stake? As shared at the beginning of this book, what is at stake is everything.

There is a response to this. We see it in Food First/The Institute for Food and Development Policy, operating out of Oakland, California; we see it in La Via Campesina, the world's largest movement of peasant farmers, in the Southern Hemisphere; we see it in the Landless Workers' Movement in Brazil, which uses legal machinations and their own bodies to take on inequalities and reclaim unused farming lands; we see it in Will Allen's transformative Growing Power, working to change urban agriculture; we see it in farmers the world over saving heirloom seeds and creating seed banks open to any; we see it in the growing resistance to Monsanto the world over, despite the latter's revolving door of power with the US (and increasingly global) government. We see it in the proliferation of CSAs, farmers' markets, and the bourgeoning interest in sustainable agriculture in the United States. We also see that some progressive activists, theologians, campaigners, organizers, and foodies in various religions realize what is at stake as well. What unites all of the above is the desire to take control of their bodies, communities, and soils, to disengage from the perceived ills of industrial agriculture, and to create ways of farming and eating and living that are just, equitable, healthy, resilient, and democratic.

This brings me back to the central theme of this book: concern about food issues and environmental health is increasingly shaping religion in the United States. This is occurring at all levels where through lived networks members and institutions of religions are shaping their respective teachings and practices about locality, health, and justice. This growth is occurring not just within religions but across them, so interfaith concerns about and involvement in ecological agrarian issues are furthering the influence and impact of the religious agrarian worldview. This is seen in the online publication and open sharing of GreenFaith's interfaith teaching book, *Repairing Eden: Sustainable, Healthy Food Opportunities for Religious Institutions* (Kennealy 2009).

If my analysis and argument is correct, then the following sensibilities shared by Farmer D—his interpretation and understanding of his own tradition and the iterative shaping of religion and farming—will continue to gain adherents. When asked why he thought religion is beginning to take seriously and thus engage with sustainable agriculture concerns (especially Jewish religion), Joffe responded by saying, "It always should have been on board in my opinion. It's always been in alignment with it. They're complimentary." For Farmer D, this alignment is "like an evolution, really. And if you're at all spiritual, you have a connection to religion and your humility and appreciation, farming is, that's farming. That's food and farming. They're so intertwined. I think religion grew out of agriculture in a lot of ways." For him and many others, sustainable agriculture, food, place, and religion combine to create an intertwined religious agrarianism that invests the local, health, and justice with political, social, and religio-spiritual meanings.

Although religious environmentalism is a recent phenomenon, it is increasing in influence because it "embodies a predominately ethics-based environmentalism, a stance which seems to be unique within the broader environmental movement" (Smith and Pulver 2009: 169). This movement of religion into environmental ethics finds some of its most vibrant expression and most committed engagement in environmental issues regarding sustainable agriculture. In this way, the translation of religious environmentalist values into embodied practice, grounded within the places of sustainable farms and farmers' markets spread throughout the United States, is becoming an emergent phenomenon in North American religions. One thing became clear to me as I undertook my own ethnographic and research trajectory in this project: a strong emphasis on caring for the Earth is indeed already flourishing within lived networks of religious agrarians and their concerns for locality, health, and justice.

Furthermore, if "the collective and ethical dimensions of narrative come together most commonly and perhaps most powerfully in religion" (Peterson 2001: 19), then narrative ethics based on relationships to place and locality can be realized through religious teachings and practice. This movement is already under way in US religions, as practitioners are developing narratives and ethics (or are adopting existing ones) around sustainable food production and are couching these in religious frames. The result is a unique, emerging, and sophisticated interweaving of ecological agrarianism with religious belief and practice as has been explored and investigated in this book.

APPENDIX

The interviews I undertook with Jewish and Christian religious agrarians in the United States were informed by grounded theory methodology. This methodology maintains that, "the questioning [of research subjects] allows interviewers to glean the ways in which research participants view their . . . world" (Strauss and Corbin 1990: 57). Therefore I located my research subjects by interacting with, observing, and interviewing the key actors involved in sustainable agriculture issues in my research communities.

Prior to my fieldwork, I generated a list of interview questions that became the basis of semi-structured and open-ended interviews with sustainable agriculture religious actors who became my research subjects. As Charmaz explains (2006), in grounded theory the data that accumulate from initial interviews is transcribed, and key indicative phrases within these data become units of analysis that are coded into categories. Such coding and subsequent generation of thematic units of analysis become the building blocks for future questions, so that in grounded theory an iterative, open-ended, constantly building understanding of a social world gradually develops. This understanding is also shaped by ongoing literature review and repeatedly revisiting transcribed interviews.

Consistent with Internal Review Board (IRB) protocol, I generated a list of research questions and submitted them, along with a standardized consent form, to the IRB board at the University of Florida. My questions and consent form were approved and given Protocol # 2009-U-437 (the original protocol expired in 2011, after which research was undertaken via College of Charleston, IRB-2013-05-14-144154). After receiving approval, I contacted Koinonia, asking if I could stay there and experience life in the community. Koinonia hosts residents for a set price each night, and I paid this fee to cover my two-week stay. I also paid for the permaculture training. David Castle's unpublished manuscript is in the public library at Koinonia and is open for public viewing. All residents at Koinonia knew I was undertaking research and that I was focusing on their sustainable agriculture initiatives in a research project. Interviews were recorded; conversations undertaken during meals or in the field were written down, as best as I could recall, after the fact. Koinonia residents also suggested other people to interview,and I asked participants in the permaculture training if they would let me interview them.

For my research with Hazon groups, I contacted the synagogues in both Atlanta and Gainesville, asking if I could visit and undertake interviews and if I could visit during a CSA drop-off. After the initial contact, leadership at both synagogues suggested I contact other members and key figures in the community. I contacted the owners of Riverview Farms, asking the same; I contacted CSA members, asking if I could commute with them to visit the farm while undertaking research. The two Hazon conferences were public events, with entrance covered by a registration fee. All public documents and presentations were used, and any specific interviews were undertaken under IRB protocol. Most of my Jewish religious agrarian interviews were recorded and transcribed, with some conversations occurring in fields or car rides or over meals written down afterward.

The following questions formed the initial basis of my semi-structured interviews:

- How long have you been a follower of your faith tradition?
- What environmental teachings does your faith advocate?
- Do these teachings inspire or impact your lifestyle? If yes: how so? If no: why not?
- What are your environmental viewpoints?
- Have your religious beliefs impacted these? If yes: how so? If no: why not?
- Has your membership in your religious community impacted these viewpoints? If yes: how so? If no: why not?
- How would you define organic farming?
- When did you start practicing/being supportive of organic farming techniques?
- What impacted or shaped this decision to support organic farming?
- Do your religious beliefs play a role in this? If yes: Which ones? Why? If no: Why not?
- How long have you been a member of this community?
- What factored into your decision to become a community member?
- What secular books or studies have impacted your environmental viewpoints?
- When you encounter another person of your faith but who belongs to a different religious community that has different environmental

views, do you enter into a conversation with them about your environmental beliefs? If yes: what sort of statements do you make to them? Why? If no: why not?

- Do your religiously based environmental beliefs influence your political beliefs and voting decisions? If yes: Which beliefs? Why? If no: Why not?
- Do you think there are pressing environmental issues that demand our concern and attention? If yes: Which ones?
- Have these concerns been impacted by your religious teachings and membership in your religious community? If no: What has shaped your environmental concerns?

All names in this book are fictitious, as per IRB protocol, except for "Farmer D," who elected to have his real name used. Names of those who gave public presentations at the Hazon Food Conferences remain their given names as these statements were made to the public at large.

NOTES

PREFACE

1. For a compelling insight into such a though experiment, see John Casti's *X-Events: The Collapse of Everything*, especially the introduction and "X-Event 2" on "When Do We Eat?," (Casti 2012: 92–108).
2. On the limits of the secular in helping to understand religious piety, see Mahmood (2005); on the construction of the secular, see Asad (2003).

CHAPTER 1

1. On this, see LeVasseur et al. (2017).
2. The American Academy of Religion has recently developed a Religion and Food Group, out of which was generated an edited volume, *Religion, Food, and Eating in North America* (Zeller et al. 2014). Readers are invited to see my chapter, "Koinonia Partners: A Demonstration Plot for Food, Fellowship, and Sustainability" (LeVasseur 2014a), for a slightly different analysis of Koinonia Farm than appears in this book.
3. http://www.resilience.org/stories/2015-06-24/how-pope-francis-s-climate-encyclical-is-liberating-the-world (accessed 12 August 2015).
4. Philosopher Roger Gottlieb (2006) argues this same conclusion, and he passionately admonishes scholars to recognize an emerging and global "religious environmentalism." See also Hitzhusen and Tucker (2013). For an astute analysis and criticism of such a religious environmentalist turn, see Tomalin (2002, 2004), and various chapters included in LeVasseur and Peterson (2017).
5. A claim with a long history within anthropology, as seen in the groundbreaking work of Roy Rappaport (1979) and Victor Turner (1967), to name two.

6. As agricultural ethicist Paul Thompson writes, "human life as we know it depends upon production. . . . More particularly, human beings do not exist without the peculiar transformations that produce food and fiber commodities" (1995: 12). This biological materialist claim is echoed by Keith Warner, who writes, "Agriculture serves as the fundamental metabolic relationship binding nature and society" (2007: 1).
7. For a good clearinghouse of facts, figures, and statistics related to animal-based industrial agriculture and the ethical, environmental, political, and ontological implications of this, see Oppenlander (2012), Simon (2013), and Andersen and Kuhn (2015). For various ways such data get framed in discussions about animals and agriculture, see LeVasseur (2016).
8. For an in-depth discussion of climate change and religious production, see the special edition of the *Journal for the American Academy of Religion* (vol. 83, no. 2) I organized in 2015.
9. Vitousek et al. (1986) estimate that on a species level, *Homo sapiens* globally consume 40 percent of the total net primary productivity of the planet.
10. Regarding living within biological limits, see Hardin (1993) and Ehrlich and Ehrlich (2008). For literature criticizing the United States (and the Global North in general), and Washington, DC consensus neoliberal economics in particular, see Mander and Goldsmith (1996) and Conca and Dabelko (1998). For a rebuttal of some of these arguments, see the work of the economist Jagdish Bhagwati (2004). It must also be noted that within the United States there is great disparity in wealth and patterns of consumption, so issues of environmental justice equally entail to the United States.
11. On the multiple point of origins of movements and discourses, especially when it comes to religion/nature interactions, see Bauman (2014).
12. For an accessible introduction to this environmental history, see Merchant (2007). See also Roderick Nash's classic *Wilderness and the American Mind* (2001), and Kirkpatrick Sale's (1993) work on the ferment that helped lead to today's American environmentalisms.
13. Regarding the social construction of religion, see both McCutcheon (1997) and Burton Mack, who writes, "Social theory . . . regards religion as a human creation on a par with the other systems of signs and patterns of practices that humans have invented to structure their societies" (2001: 83). Regarding changing concepts and the power-laden social construction of what counts as sacred, see Chidester and Linenthal (1995) and Anttonen (2000). For an easily accessible look at the creation of the concept of "religion" itself, see Nongbri (2013), and at the uneven power dynamics that

have resulted from the Christian-centric view implied in such a term, see Asad (1993), among others. I do not agree with the implied logical conclusions of some of these trains of thought when taken to their extreme, especially that of McCutcheon, who suggests that "religion" is a dead-end category that might not be worth studying. The quasi–navel gazing about the constructed discourses, categories, and domains about religion that have developed within academia and what these have to say about academics is worthy of intellectual discussion; given the dual reality of climate change, and that most people outside of religious studies scholar agree that there is some "thing" called "religion" that heavily influences their lives, leads me to personally dissociate myself from the possible abuses of this line of thinking (although I am in agreement with much of what McCutcheon writes and argues). For a cogent summary and analysis of precisely this debate, and the articulate defense of a critical realist position that I hold, see Schilbrack (2014), especially chapter 4. Last, for an astute counterargument that there is merit in questing for a workable and essentialist definition of religion, see Schaffalitzky de Muckadell (2014).

14. On this, see especially Bron Taylor (2011) and his call for a more robust social scientific approach to investigating religion–nature interactions, a call I thoroughly second. See also Wexler (2016) and Taylor et al. (2016).
15. For a summary of how I carried out my research, the questions I asked, and how I found research participants, please see the Appendix.
16. The trope of "lived religion" gained standing in the field of religious studies with the work of Robert Orsi, and the first systematic theoretical thinking with of this approach came with the 1997 publication of David Hall's edited volume *Lived Religion in America*. US religious history expert Charles Lippy (1994) uses the term "popular religion," although the phrase is synonymous with "lived religion" and the latter term is the most popular one used today.
17. Also see Orsi (2003).
18. As David Harvey points out, "Values and goals (what we might call the 'teleological' as well as the 'Utopian' moment of reflexive thought), are not imposed as universal abstractions from outside but arrived at through a *living process* (including intellectual enquiry) embedded in forms of praxis and plays of power attaching to the exploration of this or that potentiality (in ourselves as well as in the world we inhabit)" (1996: 56; emphasis added). This means that religious agrarians arrive at their values by a living process embedded in religious and sustainable agriculture praxis (and their fusion), while at the same time they attempt to articulate a

liberative potentiality, freeing modern food systems from the tyranny of industrial capitalist agriculture.

19. As Tweed himself claims, "Scholars function within a network of social exchange and in a particular geographical location, and in their work they use collectively constructed professional standards. They stand in a built environment, a social network, and a professional community. In this view, theories of religion are sightings from particular geographical and social sites whereby scholars construct meaning, using categories and criteria they inherit, revise, and create" (2006: 18).
20. For a critical, self-reflexive exploration of actor network theory and thus a network approach, see Hassard and Law (1999).
21. Arjun Appadurai similarly writes about the various global flows of ideas, ethnicities, media imagery, technologies, and finances that constitute modernity (what he terms ethno-, media-, techno-, finance-, and ideoscapes) (1996). To his list I add ecological flows, or what I term "ecoscapes" (LeVasseur 2015a).
22. Acts 2:44, 45; Acts 4:32, 33: "And all who believed were together and had all things in common; and they sold their possessions and goods and distributed them to all, as any had need. . . . Now the company of those who believed were of one heart and soul, and no one said that any of the things which he possessed was his own, but they had everything in common. And with great power the apostles gave their testimony to the resurrection of the Lord Jesus, and great grace was upon them all" (quoted in Lee 1971: 25).
23. The first was in December 2009 in Monterey, California; the second was in December, 2013 at the Isabella Freedman Retreat Center in Connecticut.
24. I recognize that it is problematic to speak about some form of an essential, hegemonic tradition, like "Christianity" or "Judaism." I acknowledge that religions are plural, contested, vibrant, amorphous, polysemous, changing and a construct of those people who generate, maintain, and shape the practices, teachings, doctrines, rituals, and worldviews of the religions they inhabit. As Robert Bellah explains, when reduced to the level of the individual, religions are made of "individual believers who are never exactly the same as any other believer even in their own faith" (2011: 604). Thus, I speak and write about tiny subsections of larger fault lines within contemporary US Christianity and Judaism, with ripple effects outward to global Christianities and Judaisms, especially in regard to sustainable agricultural issues. This is the reality religious studies scholars who undertake ethnography grapple with, and words do not adequately

capture the complex, contested reality of religious lives for those who do things in the name of religion.

25. This applies to analyzing agrarian/sustainable agriculture issues at all global levels—whether in the US South, Africa, India, East Asia—and traditional ecological knowledge embodied in indigenous forms of farming, all of which are informed by power-laden concepts of gender and community; and narratives of ethnicity, belonging, and concepts of the ecological imagination (Gadgil and Guha 1995; Posey 1999; Agrawal and Gibson 2001; LeVasseur et al. 2017).

26. For a sustained, articulate examination of the history of the interaction between religion and politics in the United States, see Wuthnow (1995). For an interesting look at the cleavage in US values and politics that is in line with Wuthnow's analysis, see the work of cognitive linguist George Lakoff (2004). His analysis points to the difficulty of changing values in individuals and groups. Regarding religious values and politics, Dionne and Diiulio point out, "The best available data suggest that religious organizations and 'relationships related to religion' are clearly the major forces mobilizing volunteers in America" (2000: 5). Many of the religious volunteers they researched are involved in political activities, on both "sides" of the political spectrum. As my research suggests, some religious volunteers are becoming actively involved in sustainable agriculture networks and politics.

27. Similarly, we must be aware of the impact that ideologies have on people's values and behaviors, especially those that adhere to both religion and environmental issues. See Sunderlin for five reasons ideology is so important (2002: 3–4)

28. It needs to be pointed out that quite a few industrial farmers are contracted to grow food in certain ways on their property for corporations, where these farmers might not want to produce plant- or animal-based commodities via the industrial model. Escaping these contracts, and finding economically viable alternatives, is a key issue for any sort of transition to sustainable agriculture on US farmlands. For an engaging perspective that views the private property-based modern market as a functional religious system (and one that is inherently destructive to other religions, and human and nonhuman communities), see Loy (1997).

29. Herman Daly (2010) makes such an argument from within the discipline of environmental economics. See also Paul Hawken (1994) and Joshua Farley (2010).

30. On how congregational messages may inform individual environmental concern, see Djupe and Hunt (2009).
31. For more on theoretical approaches that analyze the gap between professed values and behavior in regard to environmental issues, see Kollmuss and Agyeman (2002). On the interplay between values, individuals, organizations, and system structure and behavior, see Senge (1990).
32. As Dave Horton argues, "Food is hugely important to the performance of green identity among environmental activists . . . food is a significant ingredient of green distinction" (2003: 70).
33. This should not be interpreted that all US citizens care about the environment or care about it equally. There are class, race, and gender differences in this concern and care. For example, Mark Stoll (2006) argues that in African American cultural groups, religion is both a central motivator and a major cause of activist concerns. Traditionally, these concerns have tended to be social in nature and it was not until the onset of environmental racism in the 1980s that the African American community took a more active interest in environmental concerns. Even here their approach and values are informed by their religion and tend to be divergent from the approaches and concerns of Caucasian Americans. My experience at Koinonia lends support to his analysis.
34. Jeremy Benstein argues that it is impossible to escape a "centric" worldview. As he points out, fish have pisco-centric views of the world; horses, equo-centric; and cats, felino-centric (2006: 54). Yet he points out that "the very existence of a human-caused environmental crisis argues for exceptional human capabilities, which bestow upon us an exclusive status with concomitant responsibilities" (ibid.: 54). For a similarly nuanced discussion about anthropocentrism, see Katz (2000).
35. For a nuanced critique of GMO technologies and attendant social and environmental impacts, written by a husband-and-wife team who are an organic farmer and seed/plant geneticist, see Adamchak and Ronald (2008).
36. This reflects not only the role of the US Department of Agriculture (USDA) in codifying national organic standards but also to the amorphous and contested nature of terms and tropes such as "organic" and "sustainable." For example, Wes Jackson argues that organic agriculture is a subset of sustainable agriculture, where the latter also includes natural systems agriculture, perennial polyculture, permaculture, biodynamics, and other sustainable farming systems.
37. For more in-depth analysis and introduction to these concepts and practices, see LeVasseur (2010b, 2010c, and 2010d). It must be noted that some

chemicals (pesticides, fungicides, and herbicides) are legally permitted and actively used by sustainable/organic farmers, and this is another point of contention within organic subcultures. Another issue is the creation of organic monocrops that are planted for increased yields to meet and shape market demands. It is acknowledged that some supporters and practitioners of sustainable agriculture are involved solely for economic reasons, whereas others are motivated partly by economic concerns and subscribe to the worldview commonly associated with sustainable agriculture.

38. Jackson founded the institute to realize his stated goal: "I think agriculture needs a technical fix—a bio-technical fix . . . This bio-technical fix would be based on mixed perennial seed-producing plants" (1985: 3). See http://www.landinstitute.org.

39. He also writes, "Soil is a placenta or matrix, a living organism which is larger than the life it supports, a tough elastic membrane which has given rise to many life forms . . . before their decline and demise. But it is itself now dying. It is a death that is utterly senseless, and portends our own. In nature the wounded placenta heals through plant succession; enterprising species cover wounds quickly. The human agricultural enterprise and all of civilization has depended upon fighting that succession" (Jackson 1985: 10). These sentiments about the soil being alive, of it being its own matrix imbued with life force, is a central tenet of sustainable agriculture and of contemporary agrarian philosophies. This premise (the soil is alive and its health equals the health of human bodies and societies) motivates farm practices (such as applying compost or planting cover crops that become green manure, or applying only chemicals that are seen as being nontoxic to soil microbes and organisms) that attempt to sustain and build the fertility and life of the soil.

40. Jackson also tellingly writes, "Anything as important as an ethic cannot be written but must evolve in the mind of a thinking community" (1985: 73). This points to the findings of my research and my argument: communities of practice enable sustainable agricultural values to be embodied and put into practice in a specific, geographically bounded place, and as this book makes clear, religious agrarian communities are at the vanguard of evolving such ethics.

41. He also notes that, "our motivations are linked to ethical and religious worldviews" (Fick 2008: 2).

42. In his words: "I have first placed agricultural sustainability in the dual context of the natural environment and human culture, both of which determine and direct the processes of agriculture . . . the way we eat and

the way we think about eating is a cultural phenomenon. It is a part of the culture of *agri-culture*. At the same time, biological and physical conditions limit what is possible for agriculture to produce. We need to be aware of both culture and nature if we are to understand the basic parts of our food systems" (Fick 2008: 5; emphasis added).

43. These tenets of sustainable agriculture include: "soil must be respected and protected"; "Adequate natural resources must be managed in locally appropriate ways"; the humane treatment of animals and the use of animals for traction power; "Agriculture must be viewed and managed in a holistic manner"; and the recognition that "Farming must offer an attractive lifestyle and a means of learning how to farm so that future generations will become farmers" (Fick 2008: 179–80).

44. In regard to "mining" the Hebrew Bible for passages and messages that speak to or evidence environmental concerns, biblical scholar Norm Habel writes, "After exploring ecology and the Bible for more than twenty years, I have come to the conclusion that . . . [t]he Bible, as a major force in the Western Christian tradition, is not obviously green and has certainly not been read by Christians—whether scholars or laity—as a work that immediately connects us personally with nature. In the past, the Bible has generally not made Christians green in their attitude to creation—just the opposite!" (2009: xvi). Although Davis's reading of scripture from an agrarian lens holds some validity, as the authors of scripture were not the "Western Christians" of whom Habel writes, it is still a problematic leap á la Tomalin to maintain they had environmental agrarian concerns. This insight holds for all contemporary religious agrarians, whether Jewish or Christian or from some other tradition. In regard to Tomalin's and Fick's appeals to scripture to find evidence of how biblical peoples dealt with the challenges of agriculture, it must be pointed out that if these challenges (pests, drought, blights, and more) were present back then, then as Jackson writes there might be an inherent problem with till agriculture itself. A critique of the general historical enterprise of agriculture is addressed later in this chapter.

45. For an in-depth look at holism, organicism, and the rise of a Western ecological worldview (and how this affects the current understanding and quest for sustainability), see Worster (1996). For a primer on resilience thinking and how this relates to managing biosocial systems, see Walker and Salt (2006).

46. Beus and Dunlap (1990) situate the cleavage of agricultural worldviews, ethics, and values, and thus the practices they lead to, into two sets of competing views and ideals about meritorious farming systems.

47. Nigel Pennick shares a similar thesis as does Casey, writing, "We in the West suffer from what John Steele calls 'Geomantic Amnesia,' a condition where place is no longer recognized as having an effect upon [an] individuals' state of mind or well-being" (1988: 7).
48. For another perspective on this process of privileging a universal and uniform space over a particular, culture-bound place, see both Harvey (1990) and Hornborg (1996). For an in-depth exploration of what this does to us on individual and cultural levels, and to the unique places of the planet, see Berry (2000).
49. Yet we must be leery of the tyranny of overprivileging the local when thinking about place, especially in regard to morality and food issues. On this, see especially Peterson (2013), and Mallory (2013) on embedded power dynamics based on race and gender and how they manifest in local food networks. These issues are further explored throughout the book, and the work of these scholars bears direct relevance to my chapter on the local.
50. One weakness in Casey's work is that it focuses on the voice and ideas of Caucasian male philosophers. In part this is a reflection on the Western philosophical tradition, which is defined largely by those thinkers. However, there exist other strains in modern and postmodern philosophy that point toward the reemergence of place. One of these is the feminist care tradition, especially as it relates to animal ethics and embodiment, and thus by extension how diet and ethical care can affect human–place practices (see synergies with the rise in concern of how farm animals are treated, and even the rise of "veganic" farming in some sustainable farming subcultures—for example, Kadhigar farm in Maine) (Donovan and Adams 2007). Another strain is ecofeminism, with its call for intersubjectivity and a return of the erotic (Glazebrook 2004; Mallory 2013). Further strains advocating for a return to embodiment and the need for this return to move toward sustainable ethics and lifestyles are seen in ecowomanism, deep ecology, and the bioregional movement.
51. For one example of a fairly nuanced and balanced perspective on these criticisms, see Fromartz (2006).
52. On sales, see the Organic Trade Association's website, http://www.organicnewsroom.com/2006/05/organic_sales_continue_to_grow.html. On certified acreage, see the US Department of Agriculture: http://www.ers.usda.gov/Data/organic/.
53. A 2002 USDA survey of 210 farmers' markets found that one-third of all vendors offer certified organic produce (Fromartz 2006). This number

is dated, so the expectation is that more than a third now offer certified organic produce.

54. For a fantastic, rich exploration of indigenous lifeways in relation to sustainable agriculture written by an anthropologist, see Apffel-Marglin's *Subversive Spiritualities* (2011).

55. Bahnson's memoir offers reflections on Judaism and sustainable agriculture in his chapter on sukkot at Adamah Farm, but this minimal comparative element is undertaken within his own permaculture-infused theological exegesis on Christianity and food.

56. Norman Wirzba and Ellen Davis are active in agrarian scholarship, but their work is largely written from a theological perspective and uses as an exemplar Wendell Berry and his Christian (Protestant)-based agrarianism. Scholars focusing on Global South agrarianism deal with religion (Parajuli 2001), but they do not focus on Global North agrarianism and/or religion. Last, rural studies undertaken in departments of geography (mainly in Europe and Canada) tend to elide religion from their research programs (see Cloke and Little 1997; Haartssen et al. 2000).

57. Kim Smith offers a similar conclusion, writing that "Agrarian ideology has, if anything, become more ubiquitous as farmers seek allies and policy makers try to make sense of the economic and social problems facing rural America. Its persistence suggests that there is more to this tradition than scholars have recognized: that it expresses values many Americans embrace, that it offers a useful way to understand the problems of farmers and the rural community generally, and that it may even point us toward solutions" (2003: 4).

58. I recognize that this history was largely about male, Caucasian landholding farmers.

59. Bizzell writes that in the late 1800s, "The balance of power was clearly on the side of the wealthy combines. The great corporations which rapidly developed into trusts used the machinery of government to accomplish their ends" (1926: 157). The same charge against government-backed corporate control of farming is a prevalent theme within today's agrarianisms. This gives evidence that there are indeed parts of a consistent lineage between the older concerns of agrarianism and the more contemporary concerns of environmental and religious agrarianisms, at least in critiques of concentrated wealth and power in a society.

60. It must also be noted that agrarianism in the United States has often been a regional movement. This is most clearly seen in the book *I'll Take My*

Stand: The South and the Agrarian Tradition (Twelve Southerners 1962). This regional agrarian movement was partly a response to the growing mechanization of agriculture and a changing political landscape (regional and national), and it was influenced by a post–Civil War culture of white Southern gentleman farmers who tended to have aristocratic and paternal control over regional politics. There was also an underlying religious critique of industrializing society present in some of the Southern agrarians of *I'll Take My Stand*, leading to an embodied tension between Calvinist depravity, on one side, and the hope of cementing God's bounty through working piously in agricultural fields on the other (Thompson 1994).

61. For a summary of a general ecological hermeneutic approach to the Hebrew Bible and New Testament (that can be applied to texts from other traditions), see Habel and Trudinger (2008).
62. For a volume exploring religious environmentalist concerns in North American Christian communities and places of worship, see McDuff (2013).
63. I speak here of an explicitly US religious agrarianism that pulls on a dominant Christian tradition. My research shows that Jews and Christians hold a variety of theocentric worldviews based on various understandings of God as conceived and developed within Abrahamic traditions.
64. J. R. McNeill even writes that there is indeed *Something New under the Sun* (2000), given this planetary impact. In terms of religious concern about the environment and how this informs food issues, I attended a two-day workshop in October 2012, organized by GreenFaith (http://greenfaith.org/) and sponsored by fifteen religious and secular groups, held at Myers Park Baptist Church in Charlotte, NC, where Reverend Amy Brooks pointedly stated during an opening "Earth Ritual" that those present were "creating something new" in religions. The ritual evocatively stated that we were gathering to recognize "this sacred Earth we call home" so that we may cultivate "care and commitment for our earth home." The executive director of GreenFaith, Fletcher Harper, asked that those present "pray for a sense of unity in acting for protection of the Earth . . . to live in a way that blesses the Earth rather than degrading it." The leading evangelical voice in Creation Care, Richard Cizik, pointed out that religious eco-reformers have "borrowed most of our energy from the environmental movement of the 1970s," and that in religion overall he has noticed a "rising tide of interest, commitment, and awareness" in environmental issues—notice this rise is from the 1970s onward, and even then, most strongly in the past fifteen years.

CHAPTER 2

1. Millard and Linda Fuller, the official founders of Habitat, had moved to Koinonia and with Jordan started building affordable housing for poverty-stricken African American neighbors on the property. This became the idea behind Habitat, which Fuller started after moving to Americus in 1976. Thus Americus is a popular destination for both church groups and international tourists seeking to learn more about and to honor Habitat for Humanity; however, it is not well known that Habitat was equally the brainchild of Jordan, as most associate the founding only with Fuller.
2. For in-depth investigations into this aspect of a racialized, religiously sanctioned Southern agricultural aristocracy, see Isaac (1999) and Wilson (1980). For these struggles within an emergent Southern, Caucasian, agrarian intellectual tradition, most famously encountered in the work of the Twelve Southerners and their manifesto *I'll Take My Stand* (2006), see Murphy (2001).
3. It could be argued that Jordan is a theologically informed ecological progenitor to Casey's "return of place." For example, "The essence of the incarnation and the resurrection, Clarence said, was that man has to deal with God in the flesh" (Lee 1971: 186). For Jordan, this Jesus-of-the-flesh was this-worldly, was humble, got his hands dirty in the soil and local politics, and taught his followers to manifest love in this world, this place, in this community—now—by working and living together off the land.
4. For a sustained analysis of Jordan's life, the founding and history of Koinonia, and the reaction of local whites and African Americans, see Lee (1971), K'Meyer (1997), and Fuller (2003). Jordan's interpretation of the New Testament, known as the Cotton Patch Gospels, are also worth reading because they give insights into his Southern agrarian theological beliefs and teachings and are exemplary examples of unique US Protestant interpretations of the Bible (1993, 2005).
5. It is unclear in my research if Jordan was influenced by structural changes in agriculture triggered by the Agricultural Adjustment Act of 1933. While he was in conversation with emerging agricultural research courtesy of his time at Georgia State, and would have witnessed the impact of New Deal laws on farming and farm commodities, it is clear that more than anything he was inspired to begin Koinonia for the theological reasons outlined here.
6. For an updated version of a "theology of the soil" written by a Christian agronomist and soil toxicologist, see Hall (2002). See Levernier and

White for an exploration of continued poverty in Georgia's Plantation Belt, and how it is built on "region-dependent cultural characteristics that were established and institutionalized in the Antebellum South" (1998: 49), including especially residues of both sharecropping and race, as well on the recruitment of agricultural laborers by Northern industry (ibid.: 52).

7. He also lobbied his neighbors that "You ought to spend at least as much trying to help house your poor brothers whom you have seen as you do trying to house God whom you have never seen" (Lee 1971: 222).

8. Much of the information for the post-Jordan 1970s, 1980s, and volatile 1990s comes from Koinonia resident David Castle (2007), whose self-authored and unpublished memoir and summary of these years is available for public reading in Koinonia's library. Castle and his wife moved to Koinonia in 1988, and he later served on the board of directors. His passing in 2008 was a loss for the community, and his widow subsequently left the community in 2009.

9. In 2008 the community received the Community of Christ International Peace Award.

10. For an example of this strain of Protestant theocentric and "Creator-centric" environmental ethic of wonder, see DeWitt (2000) and Gustafson (1996).

11. For example, in an exploration of the rise of US agrarianism following the Civil War and running through the 1920s, W. B. Bizzell shares that, "Class consciousness grew rapidly in almost every section of the United States immediately after 1870 [because capitalists] adopted the policy of the combine, both for financial profits and for increasing economic and political power. The farmers finally realized that their individualism had made them ineffective as an influence in politics and in the business relations affecting their own interests," for "The balance of power was clearly on the side of the wealthy combines" (1926: 157). The dawning realization that corporations in the latter decades of the 1800s were beginning to dominate farming markets led to the formation of agrarian-based political parties like the Grange, the Farmers' Alliance, the People's Party, and the Farm Labor Party.

12. Koinonia built a new cafeteria and dormitory that replaced this old one. This is the community's new dining hall, guest house, and library. In 2014, the community completed building the Partnership House, an eight-bedroom house for guests and interns.

13. For relations with Bruderhof, Hutterite, and Society of Brothers, see Lee (1971).

14. Unless otherwise stated, this and all other names in this book of those I interviewed are fictitious except for Daron "Farmer D" Joffe.
15. This and subsequent quotes are from an email correspondence received July 7, 2009.
16. This and all remaining quotes are from an email correspondence on 14 July 14, 2009.
17. I attended a permaculture training in February 2010 where Everett was a facilitator. My research findings from this training are found in chapters 4 through 6.
18. Permaculture was developed by Mollison and David Holmgren and became the basis of an intentional ecovillage they founded in Australia called Crystal Waters. Mollison's name is more often associated with permaculture than is Holmgren's, although both actively write and lead workshops on permaculture today.
19. For an excellent overview of permaculture, especially the use of permaculture in intentional communities, see Lockyer and Veteto (2013).
20. A cursory summary of this meeting is available in the *Koinonia Farm Quarterly* (Fall 2008: 5).
21. See "Koinonia Farm: A Community of Biodiversity" (*Koinonia Farm Quarterly* 2009: 5).
22. Taylor's family left Koinonia to be closer to their aging parents after the completion of the first version of this manuscript.
23. For further information, see my entry "Christian Perspectives on Food and Agricultural Ethics" in the *Encyclopedia of Food and Agricultural Ethics* (LeVasseur 2015b); for a concise and informative study of agriculture in Europe, focusing especially on the United Kingdom, and Christianity see Northcott (2015).
24. Theological and political treatises in Europe during the Reformation even made appeals to pastoral-based literature. On this, see Rodman (2011).
25. This is in contrast to Mediterranean farming regimes, which used terracing, irrigation, and other forms of regionally specific farming, compared with animal-drawn plows (Mazoyer and Roudart 2006: 260).
26. This is also the period that planted the seed of what became the modern world economy that emerged out of European power (Wallerstein 2011: 36–39).
27. The only difference being that instead of leaving some pasture for hay only, European farmers alternated cereal crops with fodder crops, where "The development of these rotations went hand in hand with the development

of raising herbivores, which supplied more animal products, draft power, and manure" (Mazoyer and Roudart 2006: 313–14). More use of animals and growing more mixed crops led to more manure, which added more fertility, leading to the incorporation of row crops, then growing crops at industrial scales, all resulting in population growth and a greater human presence in the natural world.

CHAPTER 3

1. Once again, I am cognizant of the implications of essentializing a religion, which is hard to escape given the limits of language and the written form. Rather, as Jewish studies and rabbinic scholar Michael Satlow explains, "Judaism is not going to become any single thing, and Jews will continue to struggle with their tradition, using these resources to construct meaningful Jewish life within cultures and societies that rarely are characterized by a single outlook. Judaism tomorrow will be like Judaism today and yesterday—a family of communities struggling to make sense of a common identity and tradition" (2006: 291). Or, as I overheard in the dining hall at the 2013 Food Conference, when two participants were arguing aspects of Jewish custom and doctrine: "There's no monolithic existence in Jewish life."
2. Such an analysis for why Jews vote for liberal policies is challenged by Rabbi Irving Greenberg (not to be confused with the Rabbi Greenberg I interviewed), who points out that Jews have been accepted and assimilated into the national collectivity of the states at larger and larger scales beginning in the 1950s and 1960s, as evidenced in part by a "soaring intermarriage rate" (1999: 37).
3. Hiebert (2008), Davis (2009), and Hillel (2007) provide excellent analyses of early Hebrew relationships with the land and hermeneutical analyses of early biblical teachings regarding ancient Israelite landscapes.
4. This point is convincingly argued by Gillian Feeley-Harnik in *The Lord's Table: The Meaning of Food in Early Judaism and Christianity* (1994). See especially chapter 4, "Food Symbolism in the Judaic Tradition," including the discussion of the language of food in early Judaism. See also Davis (2009) and her exploration of early Hebrew agrarianism, Comins (2007), and the following from Jeremy Benstein: "There are few elements that are as connected in people's minds and lives with Judaism and Jewish culture as food . . . all holy days and celebrations have an obligatory festive meal

at their center, most festivals have specific foods associated with them, and most important, the dietary requirements of Judaism, the laws of *kashrut*, are a central focus of traditional Jewish observance" (2006: 150).

5. For a powerful exploration of this topic, including how the term *Judeo-Christian* is extremely problematic, see Schwartz (1997).
6. For an in-depth exploration of the unique history of key players within (including Abraham Joshua Heschel and Arthur Waskow) leading groups that have shaped North American and especially US Jewish environmentalism, see Seidenberg (2005). Meanwhile, Jewish scholar Hava Tirosh-Samuelson writes of "the uneasy relationship between Judaism and ecology" that currently exists in Israel (2006: 32).
7. Or as Bauman et al. explain, religions are going through an "ecological reformation" that includes steps of reform, recovery, and outright replacement (2011: 60).
8. For more background, see Lavender and Steinberg (1995), especially their chapter "Jews and the Farming Tradition," which covers "Jewish Farming in the Americas," "Jewish Farming in Eastern Europe," and "Organized Jewish Farm Settlements in the United States."
9. For a strident criticism of how this passage has been taken out of context (biblical and legal) and been turned into a clarion call for Jewish environmental stewardship, see Elion Schwartz, "*Bal Taschit*: A Jewish Environmental Precept" (2001).
10. In 2008 the "Forward 50" named him to be one of the fifty most influential Jewish leaders in the United States.
11. For more on this, go to http://hazon.org.
12. Shmita is the biblical injunction to rest or release in cycles of seven: days (Sabbath), years (shmita), and cycles of years (the jubilee). In recognition of this, Hazon generated *The Hazon Shmita Sourcebook*, written by Yigal Deutscher, with help from Anna Hanau and Nigel Savage. The project is also supported by 7 Seeds: Envisioning Sabbatical Culture, and the Jewish Farm School—showing once again how Hazon is part of lived networks that are changing Jewish views of food and agriculture throughout North America.
13. One of the people I interviewed for this book, Rabbi Greenberg of Temple B'nai Israel in Gainesville, Florida, was not so impressed with Hazon, or at least with Hazon's self-image and narrative. In his words, "I don't really see it [i.e., Hazon's Jewish environmentalism and food lobbying] as a new strain [of Judaism]. I don't think it's really new at all. I think it is really a continuation of something that's been going on for a long time.

I think what is new is there's new language and there's a new generation, but every ten years there's a new generation. When I was a student at UF [the University of Florida] thirty-five years ago now I remember we had a Seder, a Passover Seder we did at Hillel that was vegetarian and was connected with the Earth and we did it sitting on the floor. That was thirty-five years ago, that didn't start now. And I'm sure that thirty-five years prior to that there was the same thing going on. Every generation takes these issues into their own language and into their own consciousness and reimagines these things as being new issues. And I actually don't think they're new issues. . . . I think these things were always present in the Jewish community. What's different about Hazon, I mean why is Hazon growing? Hazon is growing only because people who are involved in these issues prior were involved in secular organizations. Hazon took through their bike rides and charismatic leader put a brand name on what was already there. People ate hamburgers before McDonald's; you can't say McDonald's invented sandwiches. And I don't think Hazon invented environmentalism. They may want to think they invented it; they didn't invent it. They just gave a label that was very convenient and they branded it in such a way that people all the sudden became aware. But it's branding and marketing is more what they did than really create a movement. I think the movement was there. I think you could see this echo in previous Jewish generations."

14. See http://www.shearithisrael.com, "A Home for your faith, your family, your future."
15. http://www.jewishfarmschool.org/ (accessed January 9, 2011).
16. http://isabellafreedman.org/adamah/intro.

CHAPTER 4

1. http://theculinarytrust.org/376/ (accessed January 18, 2011). The Culinary Trust has developed under different names since 1984 and today works on preserving US culinary heritage, educating culinary professionals, and providing hunger relief to the poor. They have incorporated a sustainability program into their services and in 2010 this included an Urban Food and Farm Mini-Symposium during the Annual IACP food conference. The conference included breakout sessions on topics ranging from growing food in an urban setting; backyard chickens; urban food and social justice; local meat; connecting producers and buyers; food, culture,

and community; and growing a sustainable food city. Such topics highlight the emerging environmental agrarian concerns that are influencing many aspects of US food culture.

2. On relocalizing urban agriculture, see Cockrall-King (2012).
3. For a fascinating look at ethical relations with plants, specifically, see Hall (2011). There is growing evidence that plants actively communicate with one another and have unique forms of sentience. See http://www.medicaldaily.com/scientists-confirm-plants-talk-and-listen-each-other-communication-crucial-survival-240775 (accessed June 20, 2014).
4. For an in-depth treatment of bioregionalism, see LeVasseur (2010a).
5. As an anecdote, Knoll Farms, where my wife and I worked in 2001, is a progressive biodynamic farm in Brentwood, California. They were one of the state's earliest certified organic farms and after the federal government became involved with regulating the organic food industry, the Knolls opted to forgo certification and in 2001 changed the name of their farm to Tairwa'-Knoll Farms (http://www.knollorganics.com/).
6. They further point out that "general critiques of the broader American foodscape—on grounds of health, sustainability, equity, and corporate dominance—remind us that the rarified world of foodies is a relatively exclusive, and segregated terrain. This exclusivity has a discursive dimension: despite the embrace of 'multicultural' cuisine within an expanding gourmet foodscape, dominant foodie culture in the United States continues to feature a particular demographic—namely, white and relatively affluent—as the normative ideal" (Johnston and Baumann 2010: 15–16). Such tensions were evident at Koinonia, where white members—for various environmental, religious, and political reasons—embraced permaculture lifestyles and black members viewed this as a form of modern-day sharecropping.
7. For an insightful article that speaks about a "*rooted* cosmopolitanism" (2005: 92; emphasis in original) and that offers six critiques of bioregionalism in general, see Meredith (2005). For further criticisms of bioregionalism, see Taylor (2000).
8. He also writes, "it seems to me that the particularism of distinctive *places* fashioned by human culture's peculiar and fascinating interpenetration with all the vagaries of topography, climate, and evolving ecology that define landscapes—and the continuing existence of such *places* despite the homogenizing forces of the modern world ought to cause us to realize that one of the most insightful ways for us to think about the human past is in the form of what might be called *bioregional* histories. . . . From the

time humans located regularly-visited hunting camps and early river valley farming settlements, human places have been super-imposed on environmental settings. They still are" (Flores 1999: 44; emphasis in original). The field of religion and nature will be served by longitudinal studies of how such bioregional histories are affecting and being affected by religious values. Religious agrarian communities provide one entryway into such a long-term study of human values/practices and how these relate to bioregional landscapes and identities.

9. As translated to me by a conference participant. The Birkat prayer also reads: *Baruch Eloheinu she-achainu mishelo ut-tuvo chayinu* (Praised be our God, of whose abundance we have eaten, and by whose goodness we live).

10. http://www.freshthemovie.com/about/. From the website: "FRESH celebrates the farmers, thinkers and business people across America who are re-inventing our food system. Each has witnessed the rapid transformation of our agriculture into an industrial model, and confronted the consequences: food contamination, environmental pollution, depletion of natural resources, and morbid obesity. Forging healthier, sustainable alternatives, they offer a practical vision for a future of our food and our planet. Among several main characters, FRESH features urban farmer and activist, Will Allen, the recipient of MacArthur's 2008 Genius Award; sustainable farmer and entrepreneur, Joel Salatin, made famous by Michael Pollan's book, *The Omnivore's Dilemma*; and supermarket owner, David Ball, challenging our Wal-Mart dominated economy." This film is symptomatic of the types of imagery and narratives being propagated by the sustainable food movement, with which religious agrarians are networking and being influenced by, as seen in my experience at the Tuv Ha'Aretz pick up at CSI (accessed May 31, 2011).

11. Sixteen conference participants braved below-freezing temperatures for this tour and all had heard of permaculture. Although an extremely small, already self-selected sample size, this shows not only the growth of permaculture out of Australia but suggests there are North American Jews who have been exposed to permaculture. This speaks to the growing networks of religious agrarianism in Judaism.

12. Farmer D did not want to remain anonymous for this book project, so he did not receive a pseudonym.

13. http://www.farmerd.com/.

14. On the Center on the Environment and Jewish Life, see http://www.coejl.org/~coejlor/about/. According to their website, "COEJL seeks to expand the contemporary understanding of such Jewish values as *tikkun olam*

(repairing the world) and *tzedek* (justice) to include the protection of both people and other species from environmental degradation. COEJL seeks to extend such traditions as social action and *g'milut hasadim* (performing deeds of loving kindness) to environmental action and advocacy. And *shalom* (peace or wholeness), which is at the very core of Jewish aspirations, is in its full sense harmony in all creation." Many of these concepts and themes motivate both Jewish environmentalism broadly, and the Jewish strain/s of religious agrarianism I'm arguing for herein (accessed on May 31, 2011).

15. http://www.coejl.org/~coejlor/bloome/index.php (accessed March 12, 2011).
16. http://www.hazon.org/go.php?q=/readingroom/emailArchive/emails/Omer,_Shmita,_FairFood,_Sabbatical.html&tr=y&auid=8434202 (accessed June 1, 2011).
17. For a compelling look at female change agents working within the sustainable agriculture milieu, see Anderson (2004).
18. Which they succeeded in doing, even distributing a pamphlet, part of which states: "Our farmers aren't growing meat in a factory, they are growing a healthy ecosystem that includes healthy soil and plants, healthy animals and healthy people." This sentence captures the religious agrarian concern for health in a nutshell, outside of mentioning concerns about a healthy democracy. For more, see http://www.growandbehold.com. The pamphlet shares, "Our Story: While working on organic farms, we fell in love—with each other, and with the work of feeding people good food produced by farmers who love the work and the land. We knew that the meat raised on pasture was not only of the highest quality and flavor, but also produced in a balanced ecosystem, with care for the land and the animals . . . in 2010, we founded Grow and Behold Foods on the same principles that guided the farm where we first met: delicious, nutritious food comes from farms that treat the entire ecosystem with respect." As seen throughout this chapter, similar sentiments are expressed by others involved with Hazon, CSI, and Koinonia. Such narratives and views of holism and interpretation of environmental science that organic farming is better and healthier are ubiquitous within religious agrarianism.
19. http://www.tekumah.org (accessed June 19, 2014).
20. I have changed only her first name.
21. Robinson has since left Koinonia and lives "off the grid" with his girlfriend to lessen his impact even more.
22. Paterson and Peters subsequently began dating, got married (they have since divorced), and relocated to the Catholic Worker house in Los Angeles.

This speaks to the difficulty religious communities in the United States, whether intentional or voluntary, have in keeping members, even the most committed. It also speaks to the role of religion in helping people put their values into practice, as religious communities become incubators where people can try to live their values. As with the case of Robinson, sometimes an established community may not even be radical enough in living an alternative, more sustainable lifestyle.

23. http://jubileepartners.org/ (accessed June 3, 2011). According to the website, "Jubilee Partners is an intentional Christian service community in rural northeast Georgia. Our primary ministry is offering hospitality to refugees who have newly arrived in the US. We seek to understand and live by the radical implications of following Jesus Christ. We are deeply concerned about how to be effective peacemakers and how to promote justice and understanding among our neighbors and around the world. We look to the life and teachings of Jesus as our own starting place, but we work to build bridges with people from every kind of background."
24. http://www.grassfedcow.com.
25. http://www.grassfedcow.com (accessed June 3, 2011).
26. Personal email, June 2014.
27. http://news.lowthiandesign.com/t/ViewEmail/r/16399EE4DA448305/862D169265B7109BF6A1C87C670A6B9F (accessed June 3, 2011).
28. http://news.lowthiandesign.com/t/ViewEmail/r/AD7E18584ABA6B3F/862D169265B7109BF6A1C87C670A6B9F (accessed June 3, 2011).

CHAPTER 5

1. http://ota.com/organic/benefits/nutrition.html. The Organic Trade Association is a leading lobbyist for organic farming interests. They share that "A [recent] report . . . provides evidence that organic foods contain, on average, 25 percent higher concentration of 11 nutrients than their conventional counterparts. The report was based on estimated differences in nutrient levels across 236 comparisons of organically and conventionally grown foods. *Source*: 'New Evidence Confirms the Nutritional Superiority of Plant-Based Organic Foods,'" http://www.organic-center.org/reportfiles/5367_Nutrient_Content_SSR_FINAL_V2.pdf (accessed June 6, 2011).
2. For a personal narrative/manifesto/memoir that eschews chemicals entirely, see Fukuoka (2004), first published by Rodale Press in 1978.

3. For a sobering, insightful, powerful look at the impacts of working in slaughterhouses, see Pachirat (2011).
4. See http://www.huffingtonpost.com/2010/12/23/jews-ready-to-roll-out-ne_n_800492.html for coverage of this new seal (accessed June 10, 2011). See also http://www.zeek.net/711kashrut/ for an interview with Rabbi Morris Allen, who is spearheading the heksher (hescher) tzedek movement (accessed June 10, 2011). Last, see http://www.nytimes.com/2008/10/12/magazine/12kosher-t.html?pagewanted=1&th for an in-depth framing of "kosher wars" (accessed June 10, 2011).
5. Hazon's *Food for Thought* states that, "Within Jewish sources, there are some very vocal proponents of vegetarianism. And there are also many sources in Jewish tradition that clearly permit, and encourage, meat eating, especially for celebration. It seems clear that we and the planet would be better off if most of us ate less meat, but clearly whether you choose to eat any meat is up to you. Either way, we believe—and Jewish tradition teaches—that meat consumption is something we should do thoughtfully, with full awareness" (Savage and Stevenson 2009: 104).
6. When I began this project, scientists predicted that 2011 would see the largest "dead zone" in the Gulf of Mexico, due largely to excess nutrient run-off from industrial farms, coupled with excess amounts of flood waters entering the Gulf from the Mississippi. See http://www.huffingtonpost.com/2011/06/14/gulf-dead-zone-predicted-largest-in-history_n_877188.html?ref=fb&src=sp for more information (accessed June 15, 2011). In the subsequent years, these trends have sadly and frustratingly not reversed.
7. For an analysis of the impacts climate change will have on soil organisms, see Pritchard (2011). This analysis points to the intricately linked, self-reinforcing feedback loops that influence agriculture (a point that drives holistic approaches to farming, including those of religious agrarians).

CHAPTER 6

1. The sustainable agricultural milieu is similar to the radical environmental milieu in this diagnosis about corporate malfeasance, especially in regard to the machinations of the Monsanto Corporation. They differ in their prescriptions, although the lines blur; for example, many supporters of sustainable agriculture in Europe have taken to active civil disobedience and "monkeywrenching" as they tear up test plots for genetically modified

organisms. For more on the analysis of corporations and their perceived role in despoiling the Earth and civil democracy, see Taylor (1995: 115).

2. http://magentzedek.org (accessed June 15, 2011).
3. Notice here that the creation of Magen Tzedek is a response to lived experience. In this respect, the generation of religious values and the effort to put them into practice is not always a linear process. In some cases there is an experience or exposure to a habit/practice that shapes value formation. Often it can also be an iterative process, where sometimes values originate and lead to a change in habit/practice, which leads to a further reflection on values, and then further changes in habits/practice. The concerns religious agrarians have with the local, health, and justice are partly a response to the practice of industrial agriculture. With that said, for many I spoke with, the generation of new ways of thinking about religion and sustainable agriculture and putting religious agrarian values into practice originates very much in the realm of belief, worldviews, and ideas as I have been sharing throughout this book.
4. http://www.hazon.org/go.php?q=/readingroom/emailArchive/emails/Omer,_Shmita,_FairFood,_ Sabbatical.html&tr=y&auid=8434202 (accessed May 31, 2011).
5. These two community members were not interviewed for this book, although their passions and concerns are shared here.

CHAPTER 7

1. See http://www.grassrootsnetroots.org/articles/article_20315.cfm (accessed Friday, June 29, 2011). Dubish (2004) has made similar claims in regard to the health food movement being a religion.
2. See http://www.lakejunaluska.com/uploadedFiles/Lake_Junaluska/Upcoming_Events/Spirtual_Enrichment/Caring_for_Creation/Caring-for-Creation.pdf (accessed February 17, 2011).

REFERENCES

Ableman, Michael. (2005). *Fields of Plenty: A Farmer's Journey in Search of Real Food and the People Who Grow It.* San Francisco: Chronicle Books.

Abram, David. (1996). *The Spell of the Sensuous: Perception and Language in a More-Than Human-World.* New York: Pantheon Books.

Adamchak, Raoul, and Pamela Ronald. (2008). *Tomorrow's Table: Organic Farming, Genetics, and the Future of Food.* New York: Oxford University Press.

Adams, Kenneth. (1976). Making Land More Productive. In Ethyl Corporation, *Food for America's Future: Twelve Outstanding Authorities Discuss the Country's Ability to Feed Its Multiplying Millions* (83–97). Plainview, NY: Books for Libraries Press.

Agrawal, Arun, and Clark Gibson (eds.). (2001). *Communities and the Environment: Ethnicity, Gender, and the State in Community-Based Conservation.* New Brunswick, NJ: Rutgers University Press.

Alaimo, Stacy. (2010). *Bodily Natures: Science, Environment, and the Material Self.* Bloomington: Indiana University Press.

Andersen, Kip, and Keegan Kuhn. (2015). *The Sustainability Secret: Rethinking our Diet to Transform the World.* San Rafael, CA: Earth Aware Editions.

Anderson, Anna. (2004). *Women and Sustainable Agriculture: Interviews with 14 Agents of Change.* Jefferson, NC: McFarland & Company.

Anttonen, Veikko. (2000). Sacred. In Willi Braun and Russell McCutcheon (eds.), *Guide to the Study of Religion* (271–82). New York: Cassell.

Apffel-Marglin, Frédérique. (2011). *Subversive Spiritualities: How Rituals Enact the World.* New York: Oxford University Press.

Appadurai, Arjun. (1996). *Modernity at Large: Cultural Dimensions of Globalization.* Minneapolis: University of Minnesota Press.

Artson, Bradley Shavit. (2001). Our Covenant with Stones: A Jewish Ecology of Earth. In Martin Yaffe (ed.), *Judaism and Environmental Ethics: A Reader* (161–71). Lanham, MD: Lexington Books.

Asad, Talal. (1993). *Genealogies of Religion: Discipline and Reasons of Power in Christianity and Islam*. Baltimore, MD: Johns Hopkins University Press.

———. (2003). *Formations of the Secular: Christianity, Islam, Modernity*. Stanford, CA: Stanford University Press.

Atran, Scott. (2002). *In Gods We Trust: The Evolutionary Landscape of Religion*. New York: Oxford University Press.

Bahnson, Fred. (2006). Compost for the Kingdom: An Experiment in Gardening. *Christian Century*, 123(18), 9–10.

———. (2013). *Soil and Sacrament: A Spiritual Memoir of Food and Faith*. New York: Simon & Schuster.

Bailey, Liberty Hyde. (2009). *The Holy Earth: Toward a New Environmental Ethic*. Mineola, NY: Dover Publications.

Baker, John, and Michael Winkelman (eds.). (2010). *Supernatural as Natural: A Biocultural Approach to Religion*. Upper Saddle River, NJ: Pearson Prentice Hall.

Balfour, Lady Evelyn. (2006). *The Living Soil*. Bristol: Soil Association.

Bartelt, Pearl. (1997). American Jewish Agricultural Colonies. In Donald Pitzer (ed.), *America's Communal Utopias* (352–74). Chapel Hill: University of North Carolina Press.

Baudrillard, Jean. (1994). *Simulacra and Simulation*. Trans. Sheila Faria Glaser. Ann Arbor: University of Michigan Press.

Bauman, Whitney. (2014). *Religion and Ecology: Developing a Planetary Ethic*. New York: Columbia University Press.

Bauman, Whitney, Richard Bohannon II, and Kevin O'Brien. (2011). *Grounding Religion: A Field Guide to the Study of Religion and Ecology*. New York: Routledge.

Bellah, Robert. (2011). *Religion in Human Evolution: From the Paleolithic to the Axial Age*. Cambridge, MA: Harvard University Press.

Bellah, Robert, et al. (1985). *Habits of the Heart: Individualism and Commitment in American Life*. Berkeley: University of California Press.

Bello, Walden. (2009). *The Food Wars*. New York: Verso.

Bender, Frederic. (2003). *The Culture of Extinction: Towards a Philosophy of Deep Ecology*. Amherst, MA: Humanity Books.

Benstein, Jeremy. (2006). *The Way into Judaism and the Environment*. Woodstock, NY: Jewish Lights Publishing.

Berry, Thomas. (2006). *The Dream of the Earth*. San Francisco: Sierra Club Books.

Berry, Wendell. (1977). *The Unsettling of America: Culture and Agriculture*. New York: Avon Books.

———. (1987). *Home Economics*. San Francisco: North Point Press.

———. (1990). *What Are People For? Essays by Wendell Berry*. San Francisco: North Point Press.

———. (2000). *Life Is a Miracle: An Essay against Modern Superstition*. Berkeley: Counterpoint.

———. (2003). The Agrarian Standard. In Norman Wirzba (ed.), *The Essential Agrarian Reader: The Future of Culture, Community, and the Land* (23–33). Lexington: University Press of Kentucky.

———. (2005). *The Way of Ignorance and Other Essays*. Berkeley: Shoemaker and Hoard.

———. (2009). *Bringing it to the Table: On Farming and Food*. Berkeley: Counterpoint.

Berry, Wendell, and Wes Jackson. (2009, January 4). A 50-Year Farm Bill. *New York Times*.

Beus, C., and R. Dunlap. (1990). Conventional versus Alternative Agriculture: The Paradigmatic Roots of the Debate. *Rural Sociology*, 55(4), 590–616.

Bhagwati, Jagdish. (2004). *In Defense of Globalization*. New York: Oxford University Press.

Bingham, Rev. Canon Sally (ed.) 2009. *Love God Heal Earth*. Pittsburgh: St. Lynn's Press.

Bizzell, W. B. (1926). *The Green Rising: An Historical Survey of Agrarianism, with Special Reference to the Organized Efforts of the Farmers of the United States to Improve Their Economic and Social Status*. New York: Macmillan.

Bolduc, Arthur. (2008). One Man's Trek to Sustainable Agriculture. *Acres USA: The Voice of Eco-Agriculture*, 38, 34.

Borneman, John, and Abdellah Hammoudi. (2009). The Fieldwork Encounter, Experience, and the Making of Truth. In John Borneman and Abdellah Hammoudi (eds.), *Being There: The Fieldwork Encounter and the Making of Truth* (1–24). Berkeley: University of California Press.

Boyer, Pascal. (2001). *Religion Explained: The Evolutionary Origins of Religious Thought*. New York: Basic Books.

Brunk, Conrad, and Harold Coward. (2010). *Acceptable Genes? Religious Traditions and Genetically Modified Foods*. Albany: SUNY Press.

Brunner, Karl. (1995). Continuity and Discontinuity of Roman Agricultural Knowledge in the Early Middle Ages. In Del Sweeney (ed.), *Agriculture in the Middle Ages: Technology, Practice, and Representation* (21–40). Philadelphia: University of Pennsylvania Press.

Bryman, Alan. (2001). *Social Research Methods*. Oxford: Oxford University Press.

Callicott, J. Baird. (1994). *Earth's Insights: A Survey of Ecological Ethics from the Mediterranean Basin to the Australian Outback*. Berkeley: University of California Press.

———. (2013). *Thinking Like a Planet: The Land Ethic and the Earth Ethic*. New York: Oxford University Press.

Campbell, Bruce. (1995). Ecology versus Economics in Late Thirteenth- and Early Fourteenth-Century English Agriculture. In Del Sweeney (ed.), *Agriculture in the Middle Ages: Technology, Practice, and Representation* (76–108). Philadelphia: University of Pennsylvania Press.

Carson, Rachel. (1962). *Silent Spring*. Boston: Houghton Mifflin Harcourt.

Carton, Adrian. (2012). Food in History. In Daniel Vasey et al. (eds.), *Berkshire Encyclopedia of Sustainability, Volume 4: Natural Resources and Sustainability* (147–50). Great Barrington, MA: Berkshire Publishing.

Casey, Edward. (1996). How to Get from Space to Place in a Fairly Short Stretch of Time. In Steven Feld and Keith Basso (eds.), *Sense of Place* (13–52). Santa Fe: School of American Research Press.

———. (1998). *The Fate of Place: A Philosophical History*. Berkeley: University of California Press.

Casti, John. (2012). *X-Events: The Collapse of Everything*. New York: William Morrow.

Castle, David. (2007). *A Brief History of Koinonia: The Post-Jordan Years: 1970 to 2007*. Americus, GA: Koinonia Partners. Unpublished manuscript.

Cauley, Troy. (1935). *Agrarianism: A Program for Farmers*. Chapel Hill: University of North Carolina Press.

Cetina, Karin. (2005). Complex Global Microstructures: The New Terrorist Societies. *Theory, Culture and Society*, 22(4), 213–34.

Chapman, Audrey. (2000). Science, Religion, and the Environment. In Audrey Chapman et al. (eds.), *Consumption, Population, and Sustainability: Perspectives from Science and Religion* (3–22). Washington, DC: Island Press.

Charmaz, Kathy. (2006). *Constructing Grounded Theory: A Practical Guide through Qualitative Analysis*. London: SAGE Publications.

Chase, Malcolm. (1988). *"The People's Farm" English Radical Agrarianism 1775–1840*. New York: Clarendon Press/Oxford.

Chase, Steven. (2011). *Nature as Spiritual Practice*. Grand Rapids, MI: William B. Eerdmans.

Chidester, David, and Edward Linenthal. (1995). Introduction. In David Chidester and Edward Linenthal (eds.), *American Sacred Space* (1–42). Indianapolis: Indiana University Press.

Clapp, Jennifer, and Doris Fuchs. (2009). *Corporate Power in Global Agrifood Governance*. Cambridge: The MIT Press.

Cloke, Paul, and Jo Little. (1997). *Contested Countryside Cultures: Otherness, Marginalisation, and Rurality*. New York: Routledge.

Cockrall-King, Jennifer. (2012). *Food and the City: Urban Agriculture and the New Food Revolution*. Amherst, MA: Prometheus Books.

Comins, Mike. (2007). *A Wild Faith: Jewish Ways into Wilderness, Wilderness Ways into Judaism*. Woodstock, NY: Jewish Lights Publishing.

Conca, Ken, and Geoffrey Dabelko. (1998). *Green Planet Blues: Environmental Politics from Stockholm to Kyoto*. Boulder, CO: Westview Press.

Conkin, Paul. (2009). *A Revolution Down on the Farm: The Transformation of American Agriculture since 1929*. Lexington: University Press of Kentucky.

Counihan, Carole, and Penny Van Esterik (eds.). (2008). *Food and Culture: A Reader* (2nd ed.). New York: Routledge.

Cowdrey, Albert. (1996). *This Land, This South: An Environmental History*, rev. ed. Lexington: University Press of Kentucky.

Cronon, William. (1995). The Trouble with Wilderness, or Getting Back to the Wrong Nature. In William Cronon (ed.), *Uncommon Ground: Rethinking the Human Place in Nature* (69–90). New York: Norton.

Daly, Herman. (2010). *Ecological Economics: Principles and Applications*. Washington, DC: Island Press.

Davis, Ellen. (2009). *Scripture, Culture, and Agriculture: An Agrarian Reading of the Bible*. New York: Cambridge University Press.

Deloria, Vine, Jr. (1994). *God Is Red: A Native View of Religion*. Golden, CO: Fulcrum Publishing.

DeLoughrey, Elizabeth, and George Handley (eds.). (2011). *Postcolonial Ecologies: Literatures of the Environment*. New York: Oxford University Press.

DeWitt, Calvin. (2000). Behemoth and Batrachians in the Eye of God: Responsibility to Other Kinds in Biblical Perspective. In Dieter Hessel and Rosemary Ruether (eds.), *Christianity and Ecology: Seeking the Well-Being of Earth and Humans* (291–316). Cambridge: Center for the Study of World Religions.

———. (2007). *Earth-Wise: A Biblical Response to Environmental Issues*. Grand Rapids, MI: Faith Alive.

Dionne, E. J., Jr., and John Diiulio Jr. (2000). God and the American Experiment: An Introduction. In E. J. Dionne Jr. and John Diiulio Jr. (eds.), *What's God Got to do with the American Experiment?* (1–13). Washington, DC: Brookings Institution Press.

Djupe, Paul, and Patrick Hunt. (2009). Beyond the Lynn White Thesis: Congregational Effects on Environmental Concern. *Journal for the Scientific Study of Religion*, 48(4), 670–86.

Donahue, Brian. (2001). Reclaiming the Commons. In Eric Freyfogle (ed.), *The New Agrarianism: Land, Culture, and the Community of Life* (197–216). Washington, DC: Island Press.

Donovan, Josephine, and Carol Adams. (2007). *The Feminist Care Tradition in Animal Ethics*. New York: Columbia University Press.

Dubish, Jill. (2004). You Are What You Eat: Religious Aspects of the Health Food Movement. In Carol Delaney (ed.), *Investigating Culture: An Experimental Introduction to Anthropology* (311–19). Malden, MA: Blackwell.

Eck, Diana. (2001). *A New Religious America: How a "Christian Country" Has Become the World's Most Religiously Diverse Nation*. New York: HarperOne.

Ehrlich, Paul, and Anne Ehrlich. (2008). *The Dominant Animal: Human Evolution and the Environment*. Washington, DC: Island Press/Shearwater Books.

Emerson, R., R. Fretz, and L. Shaw. (1995). *Writing Ethnographic Fieldnotes*. Chicago: University of Chicago Press.

Fairbanks, Sandra Jane. (2010). Environmental Goodness and the Challenge of American Culture. *Ethics and the Environment*, 15(2), 79–102.

Farley, Joshua. (2010). Ecological Economics. In Richard Heinberg and Daniel Lerch (eds.), *The Post Carbon Reader: Managing the 21st Century's Sustainability Crises* (259–78). Healdsburg, CA: Watershed Media.

Faulkner, Edward. (1944). *Plowman's Folly*. Norman: University of Oklahoma Press.

Feeley-Harnik, Gillian. (1994). *The Lord's Table: The Meaning of Food in Early Judaism and Christianity*. Washington, DC: Smithsonian Institution Press.

Fick, Gary. (2008). *Food, Farming, and Faith*. Albany: SUNY Press.

Fishkoff, Sue. (2010). *Kosher Nation: Why More and More of America's Food Answers to a Higher Authority*. New York: Schocken Books.

Flores, Dan. (1999). Place: Thinking about Bioregional History. In Michael Vincent McGinnis (ed.), *Bioregionalism* (43–58). London: Routledge.

Fracchia, Charles. (1979). *Living Together Alone: The New American Monasticism*. San Francisco: Harper and Row.

Freudenberger, C. Dean. (1980). Managing the Land and Water. In Charles Lutz (ed.), *Farming the Lord's Land: Christian Perspectives on American Agriculture* (123–45). Minneapolis: Augsburg Publishing House.

Freyfogle, Eric. (2007). *Agrarianism and the Good Society: Land, Culture, Conflict, Hope*. Lexington: University Press of Kentucky.

Fromartz, Samuel. (2006). *Organic, Inc.: Natural Foods and How They Grew.* New York: Harcourt.

Fukuoka, Masanobu. (2004). *The One Straw Revolution: An Introduction to Natural Farming.* Goa: Other India Press.

Fuller, Faith (dir.). (2003). *Briars in the Cottonpatch: The Story of Koinonia Farm.* Cotton Patch Productions.

Fu-Tuan, Yi. (1974). *Topophilia: A Study of Environmental Perception, Attitudes, and Values.* Upper Saddle River, NJ: Prentice Hall.

Gadgil, Madhav, and Ramachandra Guha. (1995). *Ecology and Equity: The Use and Abuse of Nature in Contemporary India.* New York: Routledge.

Gilding, Paul. (2011). *The Great Disruption: Why the Climate Crisis Will Bring on the End of Shopping and the Birth of a New World.* New York: Bloomsbury Press.

Glacken, Clarence. (1967). *Traces on the Rhodian Shore: Nature and Culture in Western Thought from Ancient Times to the End of the Eighteenth Century.* Los Angeles: University of California Press.

Glaser, Barney, and Anselm Strauss. (1967). *The Discovery of Grounded Theory: Strategies for Qualitative Research.* London: Weidenfeld and Nicolson.

Glazebrook, Trish. (2004). Eco-logic: An Erotic of Nature. In Bruce Foltz and Robert Frodeman (eds.), *Rethinking Nature: Essays in Environmental Philosophy* (95–113). Bloomington: Indiana University Press.

Goldberg, Robert. (1991). Zion in Utah: The Clarion Colony and Jewish Agrarianism. In Moses Rischin and John Livingston (eds.), *Jews of the American West* (66–91). Detroit: Wayne State University Press.

Gottlieb, Roger. (2006). *A Greener Faith: Religious Environmentalism and Our Planet's Future.* New York: Oxford University Press.

Gould, Rebecca. (2005). *At Home in Nature: Modern Homesteading and Spiritual Practice in America.* Berkeley: University of California Press.

Graham, Mark. (2005). *Sustainable Agriculture: A Christian Ethic of Gratitude.* Cleveland: Pilgrim Press.

Green Bible: New Revised Standard Version. (2008). San Francisco: HarperOne.

Greenberg, Rabbi Irving. (1999). Jewish Denominationalism Meets the Open Society. In Marjorie Garber and Rebecca Wallkowitz (eds.), *One Nation under God? Religion and American Culture* (32–59). New York: Routledge.

Greene, Shirley. (1949). Farming as a Christian Vocation: A Roundtable Discussion. *Social Action*, 15, 21–27.

Gross, Rita. (2003). Toward a Buddhist Environmental Ethic. In Richard Foltz (ed.), *Worldviews, Religion, and the Environment: A Global Anthology* (163–71). Belmont, CA: Thomson Wadsworth.

Gustafson, James. (1996). *A Sense of the Divine: The Natural Environment From a Theocentric Perspective*. Cleveland: Pilgrim Press.

Haartssen, Tialda, et al. (eds.). (2000). *Claiming Rural Identities: Dynamics, Contexts, Policies*. Assen, Netherlands: Van Gorcum.

Habel, Norman. (2009). *An Inconvenient Text: Is a Green Reading of the Bible Possible?* Adelaide, Australia: ATF Press.

Habel, Norman, and Peter Trudinger (eds.). (2008). *Exploring Ecological Hermeneutics*. Atlanta: Society of Biblical Literature.

Hall, Matthew. (2011). *Plants as Persons: A Philosophical Botany*. Albany: SUNY Press.

Hall, Steven. (2002). Toward a Theology of Sustainable Agriculture. *Perspectives on Science and Christian Faith*, 54(2), 1–5.

Hardin, Garrett. (1993). *Living within Limits: Ecology, Economics, and Population Taboos*. New York: Oxford University Press.

Harding, Stephan. (2006). *Animate Earth: Science, Intuition and Gaia*. White River Junction, CT: Chelsea Green Publishing.

Harvey, David. (1990). *The Condition of Postmodernity*. Cambridge: Oxford Publishing.

———. (1996). *Justice, Nature and the Geography of Difference*. Oxford: Blackwell.

Harvey, Graham. (2013). *Food, Sex & Strangers: Understanding Religion as Everyday Life*. Bristol, CT: Acumen.

Hassard, John, and John Law (eds.). (1999). *Actor Network Theory and After*. Malden, MA: Blackwell.

Hawken, Paul. (1994). *The Ecology of Commerce*. New York: HarperBusiness.

Heilman, Robert. (1982). Spokesman and Seer: The Agrarian Movement and European Culture. In William Harvard and Walter Sullivan (eds.), *A Band of Prophets: The Vanderbilt Agrarians after Fifty Years* (93–115). Baton Rouge: Louisiana State University Press.

Heise, Ursula. (2008). *Sense of Place and Sense of Planet: The Environmental Imagination of the Global*. New York: Oxford University Press.

Hertsgaard, Mark. (2011). *Hot: Living through the Next Fifty Years on Earth*. Boston: Houghton Mifflin Harcourt.

Hesterman, Oran. (2011). *Fair Food: Growing a Healthy, Sustainable Food System for All*. New York: Public Affairs.

Hetherington, Kregg. (2005). *Cultivating Utopia: Organic Farmers in a Conventional Landscape*. Halifax, Nova Scotia: Fernwood Publishing.

Hiebert, Theodore. (2008). *The Yahwist's Landscape: Nature and Religion in Early Israel*. Minneapolis: Augsberg Fortress.

Hilde, Thomas C. (2009.) Agrarianism. In J. Baird Callicott and Robert Frodeman (eds.), *Encyclopedia of Environmental Ethics and Philosophy* (20–23). Detroit: Gale Cengage Learning.

Hillel, Daniel. (2007). *The Natural History of the Bible: An Environmental Exploration of the Hebrew Scriptures*. New York: Columbia University Press.

Hitzhusen, Gregory, and Mary Evelyn Tucker. (2013). The Potential of Religion for Earth Stewardship. *Frontiers in Ecology and the Environment*, 11(7), 368–76.

Holthaus, Gary. (2009). *From the Farm to the Table: What All Americans Need to Know about Agriculture*. Lexington: University Press of Kentucky.

Hornborg, Alf. (1996). Ecology as Semiotics: Outlines of a Contextualist Paradigm for Human Ecology. In Philippe Descola and Gísli Pálsson (eds.), *Nature and Society: Anthropological Perspectives* (45–62). London: Routledge.

Horton, Dave. (2003). Green Distinctions: The Performance of Identity among Environmental Activists. In Bronislaw Szerszynski, Wallace Heim, and Claire Waterton (eds.), *Nature Performed: Environment, Culture and Performance* (63–77). Oxford: Blackwell/Sociological Review.

Howard, Sir Albert. (1943). *An Agricultural Testament*. New York: Oxford University Press.

Huey, Tina Andersen. (2005). Thinking Globally, Eating Locally: Website Linking and the Performance of Solidarity in Global and Local Food Movements. *Social Movement Studies*, 4(2), 123–37.

Hurt, Douglas. (1994). *American Agriculture: A Brief History*. Ames: Iowa State University Press.

Immergut, Matthew. (2008). Adamah (Earth): Searching for and Constructing a Jewish Relationship to Nature. *Worldviews: Environment, Culture, Religion*, 12, 1–24.

Ingold, Tim. (2000). *The Perception of the Environment: Essays on Livelihood, Dwelling and Skill*. New York: Routledge.

Isaac, Rhys. (1999). *The Transformation of Virginia: 1740–1790*. Chapel Hill: University of North Carolina Press.

Jackson, Wes. (1985). *New Roots for Agriculture: New Edition*. Lincoln: University of Nebraska Press.

———. (1987). *Altars of Unhewn Stone: Science and the Earth*. San Francisco: North Point Press.

———. (1996). *Becoming Native to This Place*. Berkeley, CA: Counterpoint.

Jacobs, Mark. (2002). Jewish Environmentalism: Past Accomplishments and Future Challenges. In Hava Tirosh-Samuelson (ed.), *Judaism and Ecology:*

Created World and Revealed World (449–80). Cambridge: Harvard University Press.

Jansen, Kees, and Sietze Vellema (eds.). (2004). *Agribusiness and Society: Corporate Responses to Environmentalism, Market Opportunities, and Public Regulation*. New York: Zed Books.

Johnson, Elizabeth. (2014). *Ask the Beasts: Darwin and the God of Love*. New York: Bloomsbury.

Johnston, Josee, and Shyon Baumann. (2010). *Foodies: Democracy and Distinction in the Gourmet Foodscape*. New York: Routledge.

Jordan, Clarence. (1993). *Sermon on the Mount*. Valley Forge, PA: Judson Press.

———. (2005). *The Substance of Faith and Other Cotton Patch Sermons*. Eugene, OR: Cascade Books.

Katz, Eric. (2000). Against the Inevitability of Anthropocentrism. In Andrew Light, Eric Katz, and David Rothenberg (eds.), *Beneath the Surface: Critical Essays in the Philosophy of Deep Ecology* (17–42). Cambridge, MA: MIT Press.

Kempton, Willett, et al. (1995). *Environmental Values in American Culture*. Cambridge, MA: MIT Press.

Kennealy, Stacey. (2009). *Repairing Eden: Sustainable, Healthy Food Opportunities for Religious Institutions*. Highland Park, NJ: GreenFaith.

Kent, George. (2005). *Freedom from Want: The Human Right to Adequate Food*. Washington, DC: Georgetown University Press.

Kimbrell, Andrew (ed.). (2002). *Fatal Harvest: The Tragedy of Industrial Agriculture*. Washington, DC: Foundation for Deep Ecology/Island Press.

King, Barbara. (2007). *Evolving God: A Provocative View on the Origins of Religion*. New York: Doubleday.

King, Franklin Hiram. (2004). *Farmers of Forty Centuries: Organic Farming in China, Korea, and Japan*. Mineola, NY: Dover Publications.

Kirschenmann, Frederick. (2010). *Cultivating an Ecological Conscience: Essays from a Farmer Philosopher*. Berkeley, CA: Counterpoint.

Kirschenmann, Frederick, and George Bird. (2006). Future Potential for Organic Farming: A Question of Ethics and Productivity. In Charles Francis, Raymond Poincelot, and George Bird (eds.), *Developing and Extending Sustainable Agriculture: A New Social Contract* (307–24). New York: Haworth Food and Agricultural Products Press.

K'Meyer, Tracy. (1997). *Interracialism and Christian Community in the Postwar South: The Story of Koinonia Farm*. Charlottesville: University Press of Virginia.

Koinonia Farm Quarterly. (2008). 2,2.

———. (2009). 3,1.

Kollmuss, Anja, and Julian Agyeman. (2002). Mind the Gap: Why Do People Act Environmentally and What Are the Barriers to Pro-Environmental Behavior? *Environmental Education Research*, 8(3), 239–59.

Koster, E. P. (2009). Diversity in the Determinants of Food Choice: A Psychological Perspective. *Food Quality and Preference*, 20, 70–82.

Kunstler, James Howard. (2005). *The Long Emergency: Surviving the Converging Catastrophes of the Twenty-First Century.* New York: Atlantic Monthly Press.

Lakoff, George. (2004). *Don't Think of an Elephant! Know Your Values and Frame the Debate.* White River Junction, VT: Chelsea Green.

Lappé, Anna. (2010). *Diet for a Hot Planet: The Climate Crisis at the End of Your Fork and What You Can Do about It.* New York: Bloomsbury.

Lavender, Abraham, and Clarence Steinberg. (1995). *Jewish Farmers of the Catskills: A Century of Survival.* Gainesville: University Press of Florida.

Leaf, Murray. (2012). Green Revolution. In Daniel Vasey et al. (eds.), *Berkshire Encyclopedia of Sustainability, Volume 4: Natural Resources and Sustainability* (198–204). Great Barrington, MA: Berkshire Publishing.

Lee, Dallas. (1971). *The Cotton Patch Evidence: The Story of Clarence Jordan and the Koinonia Farm Experiment (1942–1970).* Americus, GA: Koinonia Partners.

Leopold, Aldo. (1968). *A Sand County Almanac* (1949). New York: Oxford University Press.

LeVasseur, Todd. (2010a). Bioregion. In Nevin Cohen and Paul Robbins (eds.), *Green Cities: An A-to-Z guide* [online]. Thousand Oaks, CA: SAGE.

———. (2010b). Composting. In Juliana Mansvelt and Paul Robbins (eds.), *Green Consumerism: An A-to-Z guide* [online].Thousand Oaks, CA: SAGE.

———. (2010c). Green Food. In Juliana Mansvelt and Paul Robbins (eds.), *Green Consumerism: An A-to-Z guide* [online]. Thousand Oaks, CA: SAGE.

———. (2010d). Organic Trend. In Juliana Mansvelt and Paul Robbins (eds.), *Green Consumerism: An A-to-Z guide* [online].Thousand Oaks, CA: SAGE.

———. (2014a). Koinonia Partners: A Demonstration Plot for Food, Fellowship, and Sustainability. In Benjamin Zeller, Marie Dallam, Reid Neilson, and Nora Rubel (eds.), *Religion, Food, and Eating in North America* (253–73). New York: Columbia University Press.

———. (2014b). Christian Perspectives on Food and Agricultural Ethics. In Paul Thompson and David Kaplan (eds.), *Encyclopedia of Food and Agricultural Ethics* [online]. SpringerReference.com.

———. (2015a). Is Trash Hybrid? *Green Humanities: A Journal of Ecological Thought in Literature, Philosophy & the Arts*, 1(1). http://greenhumanities.org/archives/vol1/.

———. (2015b). "The Earth is *sui generis*": Destabilizing the Climate of our Field. *Journal of the American Academy of Religion*, 83(2), 300–319.

———. (2016). Methane Dispensers or Bio-Dynamic Beings ? Polysemous Meanings of Domesticated Ruminant Bovines. *Ecozon@: European Journal of Literature, Culture, and Environment*, 7 (1), 112–27.

LeVasseur, Todd, and Anna Peterson (eds.). (2017). *Religion and Ecological Crisis: The "Lynn White Thesis"at Fifty*. New York: Routledge.

LeVasseur, Todd, Norman Wirzba, and Pramod Parajuli (eds.). (2016). *World Religions and Sustainable Agriculture*. Lexington: University Press of Kentucky.

Levernier, William, and John White. (1998). The Determinants of Poverty in Georgia's Plantation Belt: Explaining the Differences in Measured Poverty Rates. *American Journal of Economics and Sociology*, 57(1), 47–70.

Lippy, Charles. (1994). *Being Religious, American Style: A History of Popular Religiosity in the United States*. Westport, CT: Greenwood Press.

Lockyer, Joshua, and Jim Veteto (eds.). (2013). *Environmental Anthropology Engaging Ecotopia: Bioregionalism, Permaculture, and Ecovillages*. Oxford: Berghahn Books.

Logsdon, Gene. (2000). *Farming at Nature's Pace: Farming and the American Dream*. White River Junction: Chelsea Green.

———. (2007). *The Mother of All Arts: Agrarianism and the Creative Impulse*. Lexington: University Press of Kentucky.

Loy, David. (1997). The Religion of the Market. *Journal of the American Academy of Religion*, 65(2), 275–90.

Lutz, Charles (ed.). (1980). *Farming the Lord's Land: Christian Perspectives on American Agriculture*. Minneapolis, MN: Augsburg Publishing.

Lyson, Thomas. (2007). Civic Agriculture and the North American Food System. In Clare Hinrichs and Thomas Lyson (eds.), *Remaking the North American Food System: Strategies for Sustainability* (19–32). Lincoln: University of Nebraska Press.

Mack, Burton. (2001). *The Christian Myth: Origins, Logic, and Legacy*. New York: Continuum.

Mahmood, Saba. (2005). *The Politics of Piety: The Islamic Revival and the Feminist Subject*. Princeton, NJ: Princeton University Press.

Mallory, Chaone. (2013). Locating Ecofeminism in Encounters with Food and Place. *Journal of Agricultural and Environmental Ethics*, 26(1), 171–89.

Mander, Jerry and Edward Goldsmith (eds.). (1996). *The Case against the Global Economy and for a Turn toward the Local*. San Francisco: Sierra Club Books.

Manning, Richard. (2004). *Against the Grain: How Agriculture Has Hijacked Civilization*. New York: North Point Press.

Maretzki, Audrey, and Elizabeth Tuckermanty. (2007). Community Food Projects and Food System Sustainability. In Clare Hinrichs and Thomas Lyson (eds.), *Remaking the North American Food System: Strategies for Sustainability* (332–44). Lincoln: University of Nebraska Press.

Marsh, Charles. (2005). *The Beloved Community: How Faith Shapes Social Justice, from the Civil Rights Movement to Today*. New York: Basic Books.

Martin, Craig. (2012). *A Critical Introduction to the Study of Religion*. Bristol, CT: Equinox.

Mayer, Stephan F., et al. (2009). Why Is Nature Beneficial?: The Role of Connectedness to Nature. *Environment and Behavior* 41, 607–42.

Mazoyer, Marcel, and Laurence Roudart. (2006). *A History of World Agriculture: From the Neolithic Age to the Current Crisis*. New York: Monthly Review Press.

McCutcheon, Russell. (1997). *Manufacturing Religion: The Discourse on Sui Generis Religion and the Politics of Nostalgia*. New York: Oxford University Press.

McDannell, Colleen. (1995). *Material Christianity: Religion and Popular Culture in America*. New Haven, CT: Yale University Press.

McDuff, Mallory. (2013). *Natural Saints: How People of Faith Are Working to Save God's Earth*. New York: Oxford University Press.

McFague, Sallie. (2013). *Collected Readings*. Ed. David Lott. Minneapolis, MN: Fortress Press.

McFarland Taylor, Sarah. (2007). *Green Sisters: A Spiritual Ecology*. Cambridge, MA: Harvard University Press.

McNeill, J. R. (2000). *Something New Under the Sun: An Environmental History of the Twentieth-Century World*. New York: Norton.

Meadows, Donella. (2008). *Thinking in Systems: A Primer*. White River Junction, VT: Chelsea Green.

Meghani, Zahra, and Jennifer Kuzma. (2010). The "Revolving Door" between Regulatory Agencies and Industry: A Problem That Requires Reconceptualizing Objectivity. *Journal of Agricultural and Environmental Ethics*, 24, 575–99.

Merchant, Carolyn. (2007). *American Environmental History*. New York: Columbia University Press.

Meredith, Dianne. (2005). The Bioregion as a Communitarian Micro-region (and its Limitations). *Ethics Place and Environment*, 8(1), 83–94.

Minkoff-Zern, Laura-Anne, and Christy Getz. (2011). Farmworkers—The Basis and Bottom of the Food Chain. *Race, Poverty, and The Environment*, 18(1), 17–19.

Moll, Rob. (2007). The Good Shepherds: A Small but Vigorous Movement Believes That Farming Is the Preservation of the World. *Christianity Today* 51(10), 64–69.

Mollison, Bill. (1979). *Permaculture Two: Practical Design for Town and Country in Permanent Agriculture*. Tasmania, Australia: Tagari Books.

Montmarquet, James. (1989). *The Idea of Agrarianism: From Hunter-Gatherer to Agrarian Radical in Western Culture*. Moscow: University of Idaho Press.

Morton, Timothy. (2010). *The Ecological Thought*. Cambridge, MA: Harvard University Press.

———. (2013). *Hyperobjects: Philosophy and Ecology after the End of the World*. Minneapolis: University of Minnesota Press.

Murphy, Paul. (2001). *The Rebuke of History: The Southern Agrarians and American Conservative Thought*. Chapel Hill: University of North Carolina Press.

Nabhan, Gary Paul. (2008). *Renewing America's Food Traditions: Saving and Savoring the Continent's Most Endangered Foods*. White River Junction, VT: Chelsea Green.

Nash, James. (1992). *Loving Nature: Ecological Integrity and Christian Responsibility*. Nashville, TN: Abingdon Press.

Nash, Roderick. (2001). *Wilderness and the American Mind* (4th ed.). New Haven, CT: Yale University Press.

Nongbri, Brent. (2013). *Before Religion: A History of a Modern Concept*. New Haven, CT: Yale University Press.

Norberg-Hodge, Helena, Peter Goering, and John Page (eds.). (2001). *From the Ground Up: Rethinking Industrial Agriculture*. New York: Zed Books/International Society for Ecology and Culture.

Nordhaus, Ted, and Michael Shellenberger. (2009). *Break Through: Why We Can't Leave Saving the Planet to Environmentalists*. New York: Mariner Books.

Northcott, Michael. (2015). *Place, Ecology and the Sacred: The Moral Geography of Sustainable Communities*. London: Bloomsbury.

Nützenadel, Alexander, and Frank Trentmann (eds.). (2008). *Food and Globalization: Consumption, Markets and Politics in the Modern World*. New York: Berg.

Oelschlaeger, Max. (1994). *Caring for Creation: An Ecumenical Approach to the Environmental Crisis*. New Haven, CT: Yale University Press.

Oppenlander, Richard. (2012). *Comfortably Unaware: What We Choose to Eat Is Killing Us and Our Planet*. New York: Beaufort Books.

Orr, David. (2001). The Urban-Agrarian Mind. In Eric Freyfogle (ed.), *The New Agrarianism: Land, Culture, and the Community of Life* (93–110). Washington, DC: Island Press.

Orsi, Robert. (1997). Everyday Miracles. In David Hall (ed.), *Lived Religion in America: Toward a History of Practice* (3–21). Princeton, NJ: Princeton University Press.

———. (2002). *The Madonna of 115th Street: Faith and Community in Italian Harlem, 1880–1950*. New Haven, CT: Yale University Press.

———. (2003). Is the Study of Lived Religion Irrelevant to the World We Live In? Special Presidential Plenary Address, Society for the Scientific Study of Religion, Salt Lake City, November 2, 2002. *Journal for the Scientific Study of Religion*, 42(2), 169–74.

Pachirat, Timothy. (2011). *Every Twelve Seconds: Industrialized Slaughter and the Politics of Sight*. New Haven, CT: Yale University Press.

Parajuli, Pramod. (2001). No Nature Apart: Adivasi Cosmovision and Ecological Discourses in Jharkhad, India. In Philip Arnold and Ann Grodzins Gold (eds.), *Sacred Landscapes and Cultural Politics: Planting a Tree* (83–113). Aldershot, UK: Ashgate Press.

Patel, Raj. (2007). *Stuffed and Starved: Markets, Power, and the Hidden Battle for the World's Food System*. Toronto: HarperCollins.

Pennick, Nigel. (1988). *Earth Harmony: Siting and Protecting Your Home: A Practical and Spiritual Guide*. Newton Abbot, UK: David & Charles.

Peterson, Anna. (2001). *Being Human: Ethics, Environment, and Our Place in the World*. Berkeley: University of California Press.

———. (2006). Toward a Materialist Environmental Ethic. *Environmental Ethics*, 28, 375–93.

———. (2009). *Everyday Ethics and Social Change: The Education of Desire*. New York: Columbia University Press.

Peterson, Gregory. (2013). Is Eating Locally a Moral Obligation? *Journal of Agricultural and Environmental Ethics*, 26, 421–37.

Pollan, Michael. (2007). *The Omnivore's Dilemma: A Natural History of Four Meals*. New York: Penguin Books.

———. (2008). *In Defense of Food: An Eater's Manifesto*. New York: Penguin Books.

———. (2009). *Food Rules: An Eater's Manual*. New York: Penguin Books.

Posey, Darrell Addison (ed.). (1999). *Cultural and Spiritual Values of Biodiversity.* London: United Nations Environment Programme.

Price, Weston. (2006). *Nutrition and Physical Degeneration* (7th ed.). La Mesa, CA: Price-Pottenger Nutrition Foundation.

Pritchard, S. G. (2011). Soil Organisms and Global Climate Change. *Plant Pathology*, 60, 82–99.

Proctor, James, and Evan Berry. (2005). Social Science on Religion and Nature. In Bron Taylor (ed.), *The Encyclopedia of Religion and Nature* (1571–77). New York: Thoemmes Continuum.

Prothero, Stephen. (2007). *Religious Literacy: What Every American Needs to Know—And Doesn't.* New York: HarperOne.

Quinn, Daniel. (1992). *Ishmael: An Adventure of the Mind and Spirit.* New York: Bantam/Turner.

———. (1996). *The Story of B: An Adventure of the Mind and Spirit.* New York: Bantam/Turner.

Radding, Cynthia. (2005). *Landscapes of Power and Identity: Comparative Histories in the Sonoran Desert and the Forests of Amazonia from Colony to Republic.* Durham, NC: Duke University Press.

Ransom, John Crowe. (1962). Reconstructed But Unregenerate. In Twelve Southerners, *I'll Take My Stand: The South and the Agrarian Tradition* (1–27). New York: Harper and Brothers.

Rappaport, Roy. (1979). *Ecology, Meaning, and Religion.* Berkeley, CA: North Atlantic Books.

Rasmussen, Larry. (2013). *Earth-Honoring Faith: Religious Ethics in a New Key.* New York: Oxford University Press.

Rischin, Moses. (1991). The Jewish Experience in America: A View from the West. In Moses Rischin and John Livingston (eds.), *Jews of the American West* (26–47). Detroit: Wayne State University Press.

Rodale, Maria. (2010). *Organic Manifesto: How Organic Farming Can Heal Our Planet, Feed the World, and Keep Us Safe.* New York: Rodale.

Rodman, Mike. (2011). *Radical Pastoral, 1381–1594: Appropriation and the Writing of Religious Controversy.* Farnham, UK: Ashgate.

Rolston, Holmes, III. (2003). *Genes, Genesis and God: Values and Their Origins in Natural and Human History.* New York: Cambridge University Press.

Roos, Bonnie, and Alex Hunt (eds). (2010). *Postcolonial Green: Environmental Politics and World Narratives.* Charlottesville: University of Virginia Press.

Rossi, Holly Lebowitz. (2008). God in the Garden: How Conservative Evangelical Churches Are Finding Feed-the-Hungry, Save-the-Planet Meaning in Their Own Backyards. *Science & Spirit*, 19(4), 40–45.

Roszak, Theodore, et al. (eds.). (1995). *Ecopsychology: Restoring the Earth, Healing the Mind*. San Francisco: Sierra Club Books.

Ruether, Rosemary Radford. (1978). The Biblical Vision of the Ecological Crisis. *Christian Century*, 1132.

Runzo, Joseph, Nancy Martin, and Arvind Sharma (eds.). (2003). *Human Rights and Responsibilities in the World Religions*. Oxford: Oneworld Publications.

Rutba House (ed.). (2005). *School(s) for Conversion: 12 Marks of a New Monasticism*. Eugene, OR: Cascade Books.

Sack, Daniel. (2000). *Whitebread Protestants: Food and Religion in American Culture*. New York: St. Martin's.

Salatin, Joel. (2007). *Everything I Want to Do Is Illegal: War Stories from the Local Food Front*. Swoope, VA: Polyface.

Sale, Kirkpatrick. (1993). *The Green Revolution: The American Environmental Movement 1962–1992*. New York: Hill and Wang.

———. (2006). *Christopher Columbus and the Conquest of Paradise*. London: Taurus.

Sanford, Whitney. (2011). Ethics, Narrative, and Agriculture: Transforming Agricultural Practice through Ecological Imagination. *Journal of Agricultural and Environmental Ethics*, 24, 283–303.

———. (2012). *Growing Stories from India: Religion and the Fate of Agriculture*. Lexington: University Press of Kentucky.

Satlow, Michael. (2006). *Creating Judaism: History, Tradition, Practice*. New York: Columbia University Press.

Savage, Nigel. (2007). Foreword. In Rabbi Mike Comins, *A Wild Faith: Jewish Ways into Wilderness, Wilderness Ways into Judaism* (xi–xiv). Woodstock, NY: Jewish Lights Publishing.

Savage, Nigel, and Anna Stevenson. (2009). *Food for Thought: Hazon's Sourcebook on Jews, Food and Contemporary Life*. New York: Hazon.

Schaefer, Donovan. (2015). *Religious Affects: Animality, Evolution, and Power*. Durham, NC: Duke University Press.

Schaffalitzky de Muckadell, Caroline. (2014). On Essentialism and Real Definitions of Religion. *Journal of the American Academy of Religion*, 82(2), 495–520.

Scheindlin, Raymond. (1998). *A Short History of the Jewish People: From Legendary Times to Modern Statehood*. New York: Macmillan.

Scherr, Sara, and Jeffrey McNeely (eds.). (2007). *Farming with Nature: The Science and Practice of Ecoagriculture*. Washington, DC: Island Press.

Schilbrack, Kevin. (2014). *Philosophy and the Study of Religions: A Manifesto*. Hoboken, NJ: Wiley-Blackwell.

Schlosser, Eric. (2005). *Fast Food Nation: The Dark Side of the All-American Meal*. New York: Harper Perennial.

Schwartz, Elion. (2001). *Bal Taschit*: A Jewish Environmental Precept. In Martin Yaffe (ed.), *Judaism and Environmental Ethics: A Reader* (230–49). Lanham, MD: Lexington Books.

Schwarz, Rabbi Sidney. (2006). *Judaism and Justice: The Jewish Passion to Repair the World*. Woodstock, NY: Jewish Lights Publishing.

Schwartz, Regina. (1997). *The Curse of Cain: The Violent Legacy of Monotheism*. Chicago: University of Chicago Press.

Schwartz, Richard. (2002). *Judaism and Global Survival*. New York: Lantern Books.

Séguy, Jean. (1973). Religion and Agricultural Success: The Vocational Life of the French Anabaptists from the Seventeenth to the Nineteenth Centuries. *Mennonite Quarterly Review*, 47(3), 179–224.

Seidenberg, David. (2005). Jewish Environmentalism in North America. In Bron Taylor (ed.), *The Encyclopedia of Religion and Nature* (909–13). New York: Thoemmes Continuum.

Senge, Peter. (1990). *The Fifth Discipline: The Art & Practice of the Learning Organization*. New York: DoubleDay.

Shepard, Paul. (1998). *Coming Home to the Pleistocene*. Washington, DC: Island Press.

Shiva, Vandana. (1993). *Monocultures of the Mind: Perspectives on Biodiversity and Biotechnology*. Atlantic Highlands, NJ: Zed Books.

———. (2000). *Stolen Harvest: The Hijacking of the Global Food Supply*. Cambridge, MA: South End Press.

———. (2016). *Who Really Feeds the World? The Failures of Agribusiness and the Promise of Agroecology*. Berkeley: North Atlantic Books.

Shortridge, Barbara, and James Shortridge (eds.). (1998). *The Taste of American Place: A Readeron Regional and Ethnic Foods*. New York: Rowman and Littlefield.

Simon, David. (2013). *Meatonomics: How the Rigged Economics of Meat and Dairy Make You Consume Too Much—And How to Eat Better, Live Longer, and Spend Smarter*. San Francisco: Conari Press.

Sine, Tom. (2008). *The New Conspirators: Creating the Future One Mustard Seed at a Time*. Downers Grove, IL: IVP Books.

Smart, Ninian. (1998). *The World's Religions* (2nd ed.). New York: Cambridge University Press.

Smith, Angela, and Simone Pulver. (2009). Ethics-Based Environmentalism in Practice: Religious-Environmental Organization in the United States. *Worldviews: Environment, Culture, Religion*, 13(2), 145–79.

Smith, Kimberly. (2003). *Wendell Berry and the Agrarian Tradition: A Common Grace*. Lawrence: University of Kansas Press.

Spencer, Nick, Robert White, and Virginia Vroblesky. (2009). *Christianity, Climate Change, and Sustainable Living*. Peabody, MA: Hendrickson Publishers Marketing.

Steiner, Rudolph. (2003). *Agriculture: An Introductory Reader*. Forest Row, UK: Sophia Books.

Stock, Jon, Tim Otto, and Jonathan Wilson-Hartgrove. (2007). *Inhabiting the Church: Biblical Wisdom for a New Monasticism*. Eugene, OR: Cascade Books.

Stoll, Mark. (2006). Religion and African American Environmental Activism. In Dianne Glave and Mark Stoll (eds.), *"To Love the Wind and the Rain": African Americans and Environmental History* (150–63). Pittsburgh: University of Pittsburgh Press.

Strausberg, Michael, and Steven Engler. (2014). *The Routledge Handbook of Research Methods in the Study of Religion*. New York: Routledge.

Strauss, Anselm, and Juliet Corbin. (1990). *Basics of Qualitative Research: Grounded Theory Procedures and Techniques*. London: SAGE Publications.

Sullivan, Lawrence. (2000). Preface. In Christopher Key Chapple and Mary Evelyn Tucker (eds.), *Hinduism and Ecology: The Intersection of Earth, Sky, and Water* (xi–xiv). Cambridge, MA: Harvard University Press.

Sunderlin, William. (2002). *Ideology, Social Theory, and the Environment*. New York: Rowman and Littlefield.

Sutterfield, Ragan. (2008). Growing Power: The Foods Resource Bank Bridges the Rural-Urban Divide through Farming and Giving. *Sojourners Magazine*, 37(2), 40–45.

———. (2013). *Cultivating Reality: How the Soil Might Save Us*. Eugene, OR: Wipf and Stock.

Taves, Ann. (2009). *Religious Experience Reconsidered: A Building-Block Approach to the Study of Religion and Other Special Things*. Princeton, NJ: Princeton University Press.

Taylor, Bron. (1995). Popular Ecological Resistance and Radical Environmentalism. In Bron Taylor (ed.), *Ecological Resistance Movements: The Global Emergence of Radical and Popular Environmentalism* (334–54). Albany: SUNY Press.

———. (1995). Resacralizing Earth: Pagan Environmentalism and the Restoration of Turtle Island. In David Chidester and Edward Linenthal (eds.), *American Sacred Space* (97–151). Bloomington: Indiana University Press.

———. (2000). Bioregionalism: An Ethics of Loyalty to Place. *Landscape Journal*, 19(1), 50–72.

———. (2002). Diggers, Wolves, Ents, Elves and Expanding Universes: Global Bricolage and the Question of Violence within the Subcultures of Radical Environmentalism. In Jeffrey Kaplan and Helene Loow (eds.), *The Cultic Milieu: Oppositional Subcultures in an Age of Globalization* (26–74). Walnut Creek, CA: AltaMira Press.

———. (2004). A Green Future for Religion? *Futures*, 36, 991–1008.

———. (2011). Editor's Introduction: Toward a Robust Scientific Investigation of the "Religion" Variable. *Journal for the Study of Religion, Nature and Culture*, 5(3), 253–62.

Taylor, Bron, Gretel Van Wieren, and Bernard Zaleha. (2016). Lynn White Jr. and the Greening-of-Religion Hypothesis. *Conservation Biology*, doi: 10.1111/cobi.12735.

Thirsk, Joan. (1997). *Alternative Agriculture: A History from the Black Death to the Present Day*. New York: Oxford University Press.

Thompson, James. (1994). God and the Agrarians. *World & I*, 9(6), 426.

Thompson, Paul. (1995). *The Spirit of the Soil: Agriculture and Environmental Ethics*. New York: Routledge.

———. (2010). *The Agrarian Vision: Sustainability and Environmental Ethics*. Lexington: University Press of Kentucky.

Tirosh-Samuelson, Hava. (2006). Judaism. In Roger Gottlieb (ed.), *The Oxford Handbook of Religion and Ecology* (25–64). New York: Oxford University Press.

Tomalin, Emma. (2002). The Limitations of Religious Environmentalism for India. *Worldviews: Environment, Culture, and Religion*, 6(1), 12–30.

———. (2004). Bio-Divinity and Biodiversity: Perspectives on Religion and Environmental Conservation in India. *Numen*, 51(3), 265–95.

Trubek, Amy. (2008). *The Taste of Place: A Cultural Journey into Terroir*. Berkeley: University of California Press.

Tucker, Mary Evelyn. (2003). *Worldly Wonder: Religions Enter Their Ecological Phase*. Chicago: Open Court.

Tucker, Mary Evelyn, and John Grim. (2014). *Ecology and Religion*. Washington, DC: Island Press.

Turner, Victor. (1967). *The Forest of Symbols: Aspects of Ndembu Ritual*. Ithaca, NY: Cornell University Press.

Tweed, Thomas. (1997). *Our Lady of the Exile: Diasporic Religion at a Cuban Catholic Shrine in Miami*. New York: Oxford University Press.

———. (2006). *Crossing and Dwelling: A Theory of Religion*. Cambridge, MA: Harvard University Press.

Twelve Southerners. (2006). *I'll Take My Stand: The South and the Agrarian Tradition*, 75th anniversary ed. Baton Rouge: Louisiana State University Press.

Van Allen, Rodger. (1978). *American Religious Values and the Future of America*. Philadelphia: Fortress Press.

Van Deth, Jan, and Elinor Scarbrough (eds). (1995). *The Impact of Values*. Oxford: Oxford University Press.

Van Wieren, Gretel. (2013). *Restored in Earth: Christianity, Environmental Ethics, and Ecological Restoration*. Washington, DC: Georgetown University Press.

Vasquez, Manuel. (2005). Historicizing and Materializing the Study of Religion: The Contribution of Migration Studies. In Karen Leonard et al. (eds.), *Immigrant Faiths: Transforming Religious Life in America* (219–41). New York: Altamira Press.

———. (2008). Studying Religion in Motion: A Networks Approach. *Method and Theory in the Study of Religion*, 20(2), 151–84.

———. (2011). *More Than Belief: A Materialist Theory of Religion*. New York: Oxford University Press.

Visotzky, Burton. (1999). The Literature of the Rabbis. In Burton Visotzky and David Fishman (eds.), *From Mesopotamia to Modernity: Ten Introductions to Jewish History and Literature* (71–102). Boulder, CO: Westview Press.

van Wormer, Kathering, Fred Besthorn, and Thomas Keefe (eds.). (2007). *Human Behavior and the Social Environment Macro Level: Groups, Communities, and Organizations*. New York: Oxford University Press.

Vitousek, Peter, et al. (1986). Human Appropriation of the Products of Photosynthesis. *BioScience*, 36(6), 368–73.

Wald, Kenneth, and Allison Calhoun-Brown. (2007). *Religion and Politics in the United States*. New York: Rowman and Littlefield.

Walker, Brian, and David Salt. 2006. *Resilience Thinking: Sustaining Ecosystems and People in a Changing World*. Washington, DC: Island Press.

Wall, Glenda. (1995). Barriers to Individual Environmental Action: The Influence of Attitudes and Social Experiences. *Canadian Review of Sociology and Anthropology*, 32, 465–89.

Wallerstein, Immanuel. (2011). *The Modern World System I: Capitalist Agriculture and the Origins of the European World-Economy in the Sixteenth Century*. Berkeley: University of California Press.

Warner, Keith. (2007). *Agroecology in Action: Extending Alternative Agriculture through Social Networks*. Cambridge, MA: MIT Press.

Weber, Max (2003). *The Protestant Ethic and the Spirit of Capitalism*. Trans. T. Parsons. New York: Dover Publications.

Wexler, Jay. (2016). *When God Isn't Green: A World-Wide Journey to Places Where Religious Practice and Environmentalism Collide*. Boston: Beacon Press.

Whitney, Elspeth. (2013). The Lynn White Thesis: Reception and Legacy. *Environmental Ethics*, 35(3), 313–31.

White, Lynn, Jr. (1967). The Historical Roots of Our Ecologic Crisis. *Science*, 155, 1203–7.

Wilkinson, Katharine. (2012). *Between God & Green: How Evangelicals Are Cultivating a Middle Ground on Climate Change*. New York: Oxford University Press.

Wilson, Charles Reagan. (1980). *Baptized in Blood: The Religion of the Lost Cause, 1865–1920*. Athens: University of Georgia Press.

Wilson, Jonathan. (1997). *Living Faithfully in a Fragmented World: Lessons for the Church from MacIntyre's* After Virtue. Harrisburg, PA: Trinity Press.

Wilson-Hartgrove, Jonathan. (2008). *New Monasticism: What It Has to Say to Today's Church*. Grand Rapids, MI: Brazos Press.

Wink, Walter. (1999). *The Powers That Be: Theology for a New Millennium*. Easton, PA: Harmony Press.

Wirzba, Norman (ed.). (2003a). *The Essential Agrarian Reader: The Future of Culture, Community, and the Land*. Lexington,: University Press of Kentucky.

———. (2003b). *The Paradise of God: Renewing Religion in an Ecological Age*. New York: Oxford University Press.

———. (2005). Mark 10: Care for the Plot of God's Earth Given to Us along with Support of Our Local Economies. In Rutba House (ed.), *School(s) for Conversion: 12 Marks of a New Monasticism* (137–48). Eugene, OR: Cascade Books.

White, Lynn, Jr. (1967). The Historical Roots of Our Ecologic Crisis. *Science*, 155(3767), 1203–7.

Willer, Helga, et al. (eds.). (2008). *The World of Organic Agriculture: Statistics and Emerging Trends 2008*. London: Earthscan.

Worster, Donald. (1996). *Nature's Economy: A History of Ecological Ideas* (2nd ed.). New York: Cambridge University Press.

Wuthnow, Robert. (1995). Old Fissures and New Fractures in American Religious Life. In David Hackett (ed.), *Religion and American Culture: A Reader* (369–82). New York: Routledge.

Yiridoe, Emmanuel, Samuel Bonti-Ankomah, and Ralph Martin. (2005). Comparison of Consumer Perceptions and Preference toward Organic versus Conventionally Produced Foods: A Review and Update of the Literature. *Renewable Agriculture and Food Systems,* 20, 193–205.

Zeller, Benjamin, Marie Dallam, Reid Neilson, and Nora Rubel (eds.). (2014). *Religion, Food, and Eating in North America.* New York: Columbia University Press.

INDEX

agrarian/ism, 101–136, 137–162, 163–180
- history of, 32, 35–39, 110, 212–213, 215
  - Twelve Southerners, 110–112, 212–213, 214
- postagrarianism, xiii

agriculture (farming), 204
- criticisms of, 31–32, 210
- impacts of anthropogenic climate change on, xiv–xv, 3
- *See also* industrial agriculture; sustainable agriculture

Bailey, Liberty Hyde, 106–107

Berry, Wendell, xiii, 35–36, 37–38, 43, 92, 103–104, 111, 118, 120, 149

Christianity, 12, 45–76, 101–136, 137–162, 163–180, 181–198, 206–207
- Christian agrarians and sustainability, 65–66, 72–75
- history of farming and Christianity
  - in America, 71–72
  - in Europe, 67–71, 216–217
- Jubilee Partners, 129, 223
- New Monastics, 56–61
  - Schools for Conversion, 61
  - twelve marks of, 59–60
- *See also* Koinonia

ecological/environmental agrarianism, 1, 32, 36–39, 86, 87, 89, 102–104, 106–107, 111, 137, 145, 179, 209, 219–220
- *See also* Berry, Wendell; religious agrarianism; sustainable agriculture

food, 19–20
- values/habits, 17–18, 19

grounded theory, 9–11, 199

Hazon, 12, 88–91, 94–100, 101–136, 137–162, 163–180
- B'nai Israel, 12
- Congregation Shearith Israel, 12, 92–4
  - Riverview Farms, 12, 101–136
- Savage, Nigel, 88, 95–98, 166–167
- *See also* Judaism

health, xvi, 1, 137–162, 181–198, 222
- animal and soil, 150–157
- farm, 145–147
- land health, 23–24
- planetary, 158–161

health (*continued*)
physical (human), 141–144
societal, 147–150
spiritual, 138–141
hyperobjects, 195

industrial agriculture, xiii–xiv, 20–22, 43, 71, 164, 207
concentrated animal feeding operation (CAFO), 149, 204
criticisms of, xiv–xv, 14, 22–30, 64, 144–145, 148, 158, 164, 179–180, 204
Green Revolution, xvi, 21–22, 27, 43
values of, 16

Jackson, Wes, 25, 31, 40, 64, 139, 149, 179–180, 208, 209
Judaism, 12, 77–100, 101–136, 137–162, 163–180, 181–198, 206–207, 217–219
Adamah Farm, 95
bal taschit/tashchit, 87, 169, 183
berakhot, 85–86
chesed, 80
Heksher Tzedek, 154, 224
history of Jewish farming, including in North America, 81, 82–84, 217, 218
ingredients of a Jewish food ethic, 187, 218
Isabella Freedman Jewish Retreat Center, 88, 95–96
Jewish agrarians and sustainability, 80–82, 84–88, 99, 221–222
kosher (*kashrut*), 154–157, 218
lived networks, 79, 88–89, 94–98
Magen Tzedek, 154, 166, 225
mashgiach, 154
mitzvoth, 142
politics in the US, 78–80, 217
Exodus and Sinai consciousness, 79–80
justice, 78–80, 87
Rabbinic teachings, 77–100,
shochet, 114, 122, 154
shmita, 90, 116, 167–168, 218
Teva Learning Center, 95
tikkun olam, 146, 169, 183, 185
Torah (and teachings of) 84, 85, 87, 116, 122
Tuv Ha'Aretz, 93–94, 113–114, 170–172
tzedakah, 80, 87, 168, 169
justice, xvi, 1, 42, 163–180, 181–198

Koinonia, 11–12, 45–76, 173, 214–216, 220
Americus, Georgia, 11, 45–46, 214
farming practices, 48, 50, 63
organic, permaculture, 50–51, 62–63
Habitat for Humanity, 46, 49, 214
history of, 46–56, 57, 214
Jordan, Clarence, 11, 46–48, 173, 214
Cotton Patch Gospels, 50, 51
Jordan, Florence, 46, 51
pacificism, 47, 51, 53, 56–57
race relations, 46, 47, 48, 49, 50–52

Leopold, Aldo, 105–106
lived religion, 6–8, 60, 61, 91, 181–198, 205, 225
Orsi, Robert, 7–8, 205
and religious agrarianism, 7–8, 41, 56, 91, 110, 116–117, 129, 173, 179
See also religion; values

local (place), xvi, 1, 17, 28–30, 34, 64, 101–136, 188–195, 211
bioregionalism, 26, 108–110, 211, 220–221
farmers' markets, 29, 33
soil, 104–105
terroir, 29, 108–109

Network Theory, 8–9, 206
lived network (approach of), 9, 10, 56, 60, 67, 78, 94, 112, 116–117, 124, 129, 131, 149, 165, 177, 179, 181–198, 218, 221
moral geographies, 9
See also religion

religion, especially in the United States
and agriculture/food, 2–3, 5, 14, 17–19, 34–35
Christianity and Judaism, xvi, 1, 4–5, 12, 13–19, 103, 166, 181–198, 204–205, 206–207
definition of, 13, 166, 204–205
ecological reformation/religious environmentalism/and sustainability, 2–4, 9, 14, 34, 40–41, 67, 104, 108, 172–173, 181–198, 203, 210, 213, 218
and sustainable agriculture/food, 34
and US politics, 5, 12, 163–180, 207
religious agrarianism/agrarians, 101–136, 137–162, 163–180, 181–198), xvi, 1, 2, 6, 11–13, 16, 19, 32–34, 40–43, 65, 67, 69, 85, 87–88, 91, 94, 104–108, 111, 112, 114, 126, 129, 135, 137, 143, 149, 163–164, 165, 172–172, 179–180, 181, 205–206, 209, 210, 212, 221, 222, 225
values of, 16–18, 26–28, 32, 65
See also Christianity; Judaism; local (place); religion
resilience and sustainability/systems thinking, 4, 5, 10, 18, 24, 26, 188, 210

sustainable agriculture, 20, 23–30, 63–64, 131, 133, 145–147, 208–209, 209–210, 221, 222
community health and the local, 24–26
community supported agriculture, 7, 12, 33–34
criticisms of, 30–31
organic agriculture, 33, 37, 223
permaculture, 63–65, 99, 126–128, 165, 173, 216
values of, 26–28
US Department of Agriculture, 15, 144, 208, 211
organic certification, 30, 144, 211
See also Koinonia; local (place)

transcorporality, 195

United States environmentalism/s, 4–5, 18, 24, 82, 204, 208, 210

values, 13–19, 26–28, 32, 38, 79, 104, 181–197, 207, 208, 221
and practice, 8, 15–16, 26–29, 41, 91, 94, 110, 112, 117, 124, 128–129, 143, 146, 149–150, 156–157, 158, 164, 165, 170–173, 175, 177, 179, 205–206, 209, 223, 225
individualism, 14–15
theocentric, 41, 130, 158, 160, 179, 213, 215